Treating Gold Ores

by Arizona Bureau of Mines

with an introduction by Kerby Jackson

This work contains material that was originally published in 1932 by the Arizona Bureau of Mines.

Introduction

It has been decades since the Department of Interior released their important publication "Treating Gold Ores". First released in 1932, this important volume has been out of print and has been unavailable to the mining community since those days, with the exception of expensive original collector's copies and poorly produced digital editions.

It has often been said that *"gold is where you find it"*, but even beginning prospectors understand that their chances for finding something of value in the earth or in the streams of the Golden West are dramatically increased by going back to those places where gold and other minerals were once mined by our forerunners. Despite this, much of the contemporary information on local mining history that is currently available is mostly a result of mere local folklore and persistent rumors of major strikes, the details and facts of which, have long been distorted. Long gone are the old timers and with them, the days of first hand knowledge of the mines of the area and how they operated. Also long gone are most of their notes, their assay reports, their mine maps and personal scrapbooks, along with most of the surveys and reports that were performed for them by private and government geologists. Even published books such as this one are often retired to the local landfill or backyard burn pile by the descendents of those old timers and disappear at an alarming rate. Despite the fact that we live in the so-called "Information Age" where information is supposedly only the push of a button on a keyboard away, true insight into mining properties remains illusive and hard to come by, even to those of us who seek out this sort of information as if our lives depend upon it. Without this type of information readily available to the average independent miner, there is little hope that our metal mining industry will ever recover.

This important volume and others like it, are being presented in their entirety again, in the hope that the average prospector will no longer stumble through the overgrown hills and the tailing strewn creeks without being well informed enough to have a chance to succeed at his ventures.

Kerby Jackson
Josephine County, Oregon
May 2016

CONTENTS

Preface 3

Introduction 5

Classification of methods of treating gold ores 6

Smelting of gold ores 6

Milling of gold ores 8

Amalgamation 8

      Advantages and disadvantages 11

      Amalgamating practice 12

Gravity stamps 12

Retorting 16

Melting and refining of retort bullion 18

Ball mills 18

Huntington mills 18

Arrastras 19

Amalgamating pans 22

Cost of recovering gold by amalgamating methods...............22

Cyanidation 24

Gravity concentration 28

Flotation 29

ILLUSTRATIONS

Fig. 1.—Stamp battery 13

Fig. 2.—A retort 17

Fig. 3.—Arrastra 20

Fig. 4.—Amalgamating pan 21

PREFACE

Because few other metals can at this time be mined profitably and because the purchasing power of gold has so greatly risen, great interest is being taken in that metal and hundreds of people are seeking deposits of gold ore in Arizona. While many of these prospectors are experienced, a large number are "greenhorns" who have no idea how to extract the gold from any ore found. This little bulletin has been prepared to answer, in part at least, the large number of inquiries received by the Arizona Bureau of Mines from such people. The Bureau will welcome questions on subjects not covered by this publication, and they will be answered by experts, free of charge, if it is possible to do so without making laboratory tests. For such tests, a charge must be made.

No treatment plant of any kind should ever be erected, no matter how small the cost, until the ore to be treated has been completely tested. Thousands of dollars may easily be saved by spending a few hundred dollars on such tests. No successful mining company ever erects a mill until exhaustive tests have been made, and they are just as essential when a small, "rawhide" plant is to be built.

A complete ore test is a very time-consuming job; it may easily take the entire time of one man for a month. When it has been completed, however, it is possible to decide just what process or combination of processes will yield the largest profit, what the percentage of extraction will be, the approximate cost of the plant, and the probable profit per ton of ore treated, assuming that the full tonnage that the plant is designed to handle is treated each day.

Not only should ore be thoroughly tested before any considerable sum is spent on the erection of a treatment plant, but no such plant should be even considered until enough ore has been blocked out or otherwise developed to insure the operation of the plant for a long enough period to repay its cost and yield a satisfactory profit.

If no treatment plant of any considerable capacity were ever erected until sufficient ore to make the enterprise profitable is assured, and if the nature of all treatment plants were determined by careful, thorough ore tests, much of the risk connected with mining enterprises would be removed.

G. M. BUTLER.

April 1, 1932.

TREATING GOLD ORES

BY

T. G. CHAPMAN

Metallurgist, Arizona Bureau of Mines; Professor of Metallurgy
and Ore Dressing, University of Arizona.

INTRODUCTION

The Arizona Bureau of Mines has received many inquiries during
the past year regarding methods for treating gold ores. Due to the
similarity in the inquiries received and the fact that many letters
have been addressed to the Bureau by persons who evidently desire
replies in non-technical language, it was believed advisable for the
Bureau to publish a brief summary of present methods of treating
gold ores for the use of persons not familiar with technical terms.
It is obvious that a paper so written cannot cover the subjects included in more than a superficial manner, but it is hoped that, due
to the widespread interest in the treatment of gold ores at the
present time, the paper will be of some value to the persons for
whom it has been prepared.

The writer wishes to acknowledge the kind assistance of Dean
G. M. Butler, College of Mines and Engineering, University of
Arizona, in reading the manuscript.

For those people who desire more detailed descriptions of gold
milling processes, the following books are recommended: "Textbook of Ore Dressing," by Richards and Locke, published by
McGraw-Hill Book Co., New York City. This book covers the subjects of gravity stamps, amalgamation, gravity concentration, and
flotation, and costs $5.50. "Manual of Cyanidation," by Hamilton,
published by the same company, describes the various methods
used for cyaniding gold ores and costs $3.00.

It should also be noted that the United States Bureau of Mines
has recently published Information Circulars which describe gold
milling methods and costs as practiced in individual plants in the
United States, Canada, and Mexico. These papers may be obtained
free of charge by writing to the Director, U. S. Bureau of Mines,
Washington, D. C. The numbers and titles of the circulars dealing
with gold milling are as follows:

I. C. 6236—Milling Practice at the Alaska Juneau Concentrator.

I. C. 6408—Milling Methods and Costs at the Homestake Mine, Lead, S. Dak.

I. C. 6411—Milling Methods and Costs at the Spring Hill Concentrator, Helena, Mont.

I. C. 6433—Amalgamation Practice at the Porcupine United Gold Mines, Ltd., Timmins, Ont.

I. C. 6476—Milling Methods and Costs at the Argonaut Mill, Jackson, Calif.

I. C. 6541—Milling Methods and Costs of the Coniaurum Mines, Schumacher, Ont.

I. C. 6611—Small-Scale Placer Mining Methods.

CLASSIFICATION OF METHODS OF TREATING GOLD ORES

The common methods available for the recovery of gold from ores derived from lode deposits may be classified as follows:

1. Smelting
2. Amalgamation
3. Cyanidation
4. Gravity concentration
5. Flotation

One of these methods may be used alone or the methods may be combined, two or more being used in treating one ore. For example, after extracting part of the gold by amalgamation, a further recovery is sometimes profitably made by cyanidation, concentration, or flotation.

SMELTING OF GOLD ORES

Smelting methods cannot be often employed to advantage in small-scale plants, especially with gold ores. In small-scale smelting operations, the initial plant investment is heavy and the operating cost is high. A considerable tonnage of high-grade ore should, therefore, be developed before even considering small-scale smelting treatment. In order to obtain a high recovery of gold by smelting methods, it is necessary to use lead or copper ore in the furnace charge, and, furthermore, the lead or copper ore must at times be pepared for the smelting operation by a roasting and sintering preliminary treatment. For the reasons outlined, it has been customary for miners and prospectors operating on a small-scale basis to ship their ores to a custom smelting plant when the smelting method is to be used for the recovery of gold. The advantages to be gained by shipping rather than constructing a small smelting plant may be summarized as follows:

1. A large developed ore body is not a requirement.
2. Large initial plant cost, including development of considerable amounts of water, is not necessary.
3. The purchase or mining of copper or lead ores is not necessary.
4. The risk involved in the proposition as a whole, when no smelter is erected, is not as great, due to the smaller investment necessary.

The disadvantages often advanced for shipping and sale of ore to custom smelters, as compared to erecting small smelting plants, are first, saving in freight charges and, second, a greater profit. The first disadvantage is sound and, if any justification for small smelting plants exists, it is to be found in an isolated locality where the charge for shipping mine ore is prohibitive or when the ore is refractory to milling processes and when fuel flux, lead, or copper ores and water are available at costs which leave a balance of profit in the operation as a whole. The second advantage, namely, that the profit will be greater even at points not far distant from custom smelting plants, is not usually well founded. With present competition for ores by smelting plants and with the more efficient operation of large custom smelters as compared to small-size plants, it will usually figure to the seller's advantage to ship to those plants rather than consider the erection of a small plant.

In marketing gold ores to smelting plants, the seller should remember that the accurate sampling of a shipment is essential in order to avoid misunderstandings when settlement is made. The correct sampling of certain types of gold ores is not easy and at times is impossible without proper crushing equipment owing to the "spotty" character of such material. Again, in marketing gold ores, the shipper should send representative samples to all custom lead and copper smelting plants within reasonable shipping distances. The net returns from different plants will not always be approximately the same because one plant might need the waste material contained in the ore under consideration as a flux more than other plants. In such a case, a favorable smelting rate might be offered which would yield a higher net return to the seller. In summarizing the smelting treatment of gold ores, it can be stated that, for small properties within reasonable distances from custom lead and copper smelting plants, greater profit has often resulted in selling the material to the smelter rather than from the erection of expensive treatment plants, and, when using this method of disposing of gold ores, the shipper should undertake the proper sampling of the shipment, and, if proper crushing equipment is not available, should provide for a representative at the smelter during the sampling of his ore in order to prevent misunderstandings in settlements.

MILLING OF GOLD ORES

The milling of gold ores is usually preferable to shipment to custom smelters providing the ore responds to milling treatment at a reasonable cost and also providing the ore body is of sufficient size to justify the expenditure necessary for constructing and equipping the milling plant. As already indicated, there are four methods available for milling gold ores, namely, amalgamation, cyanidation, gravity concentration, and flotation, and, furthermore, as previously mentioned, these methods can often be combined to advantage.

AMALGAMATION

Amalgamation methods may be applied in various ways such as (1) gravity stamp mills which combine the crushing and amalgamating operation in a container called a stamp battery, (2) copper plates coated either with mercury or silver amalgam over which the ground ore pulp flows, (3) amalgamating pans, (4) arrastras which combine grinding of the ore with the amalgamating operation, and (5) sluices equipped with riffles, the purpose of using mercury in this connection being to catch fine gold. The last method is more applicable to the treatment of placer material than to the treatment of ore from lode deposits.

No matter what device is used in applying the amalgamation method, the basic principle is the same in all cases and it may be stated as follows: If clean, bright gold is brought in contact with clean, bright mercury, the two metals will alloy and form material called gold amalgam.

It should be noted that the surfaces of the gold and mercury must be clean and bright before the union of the two metals will take place. If the surface of the mercury is dark or tarnished, the gold, no matter how bright or clean, will not be caught by the mercury. Furthermore, if the mercury is broken up or subdivided into many small globules, the gold will not be caught. Mercury is rendered dark or coated or is broken up in the following ways: (1) It is tarnished by exposure to the air especially after being in use for a short period of time. (2) It is also tarnished and coated by uniting with base metals such as copper, lead, arsenic, or antimony. The source of these base metals is the ore treated or, in the case of copper, the amalgamating copper plate. (3) Mercury is also coated by certain minerals at times found in gold ores, notably talc. (4) Mercury is broken up into small globules by grease or oil derived from materials used in the mine or mill.

The methods ordinarily used to keep the surface of the mercury bright and clean are: (1) Proper care in preventing grease or lubricating materials from coming in contact with the mercury and the use of lye, soda ash, or lime for cutting grease if present. (2) The use of a very small amount of sodium cyanide in removing the tarnish or stain. (3) The addition of a minute amount of metallic sodium to the mercury. (4) The use of common salt.

In using sodium cyanide for the removal of stains from copper amalgamating plates, it should be remembered that this substance is very poisonous, even in minute quantities, and, furthermore, cyanide dissolves gold. A piece of sodium cyanide about one inch in diameter, if held in tongs and placed just above the stain for a few seconds with the ore pulp flowing over the plate, is sufficient to remove copper stains, and the use of excessive amounts of cyanide should be avoided.

Although the addition of sodium to mercury has, in exceptional cases, proved of benefit in keeping the surface of mercury in good condition, its use is not recommended in general as it causes base metals to enter the amalgam, which, after a short period of time, oxidize and cause the mercury to tarnish. In this way, it gives more trouble than if it were omitted. If sodium is found to be beneficial, it should be used in small amounts and with caution. Since sodium amalgam will be found difficult to purchase in good condition, due to the fact that it loses strength rapidly, two methods are given for its preparation, as follows:

Sodium amalgam is usually made by adding metallic sodium directly to mercury. The sodium should be cut into pieces about the size of the head of a common pin and these pieces added one by one to mercury until the sodium amalgam formed will just adhere to a clean nail. If the sodium does not readily react with the mercury, wet the end of a glass rod with water and touch the sodium resting on the surface of the mercury with the glass rod. The amalgam will then form. The amount of sodium necessary to add will, of course, depend upon the amount of mercury used and also on the condition of the mercury. Richards and Locke[1] state that one part of sodium to 2,000 parts of mercury was used with good results in the treatment of Nova Scotia gold ores containing arsenic and talcose slate. The writer ordinarily uses from one to four parts of sodium to 2,000 parts of mercury and has found the "nail" test rather than a definite weighed amount of sodium to be the best rule for its use.

Since metallic sodium is not usually available in the field, another method which utilizes common salt and the electric current of an automobile battery for the making of sodium amalgam is described In this latter method, mercury is placed in a glass container and covered with a solution of three to five ounces of common salt per quart of water. A wire from the positive pole of the battery is suspended in the salt solution and the negative wire is inserted into the mercury. A hollow glass rod placed with one end immersed in the mercury before adding the salt solution will serve to insulate the negative wire from the salt solution. The electrolysis of the salt solution results in the formation of sodium amalgam which is tested for strength by immersing a clean iron nail and noting when the amalgam is just active enough to adhere to the nail. Electrolysis for ten to fifteen minutes is usually sufficient for a few ounces of mercury, and correspondingly greater periods of time are required for large amounts.

The use of common salt for eliminating trouble caused by talcose materials has been suggested at times although no definite examples have been cited to the writer's knowledge. Salt, when used for such a purpose, is usually added intermittently, a shovelful being added to the ore pulp when found necessary.

Having considered the fact that mercury used for amalgamating gold must be clean and bright and having described the various ways by which mercury becomes tarnished, coated, or subdivided into globules and the methods used in practice for keeping the mercury in good condition, the condition of the gold for successful recovery will next be described.

---

[1] Richards, R. H. and Locke, C. E., Textbook of Ore Dressing, McGraw-Hill Book Co., New York City.

Coarse gold, in the free condition, is usually clean and bright especially after grinding of the ore, and, when in this condition, is readily recovered by amalgamation. Gold when present under the following conditions, however, cannot be amalgamated by ordinary methods.

1. Gold attached to or enclosed in quartz.
2. Gold contained in base metal sulphides, notably iron sulphide or pyrite.
3. So-called "rusty gold" which is believed to be free gold coated with iron oxide.

A fourth form of gold which is difficult to recover by amalgamation is finely divided gold. In mentioning coarse and fine gold, no definite size can be given, but any particle of gold readily visible to the naked eye, without straining the vision, might be roughly termed coarse gold. While fine gold is difficult to amalgamate, the operation may be assisted by impinging the gold on the surface of mercury. The impinging action is usually obtained by placing amalgamating plates within the stamp battery and thereby utilizing the splash, caused by the dropping of the stamps, to throw the ore pulp containing the gold onto the surface of the mercury or by breaking up the amalgamating plate that follows the stamp battery into a number of short sections, arranged as steps, the drop from step to step being about half an inch. As the ore pulp drops a distance of half an inch from step to step, the gold particles impinge upon the mercury coating of the plate, a condition which increases the amount of the finely divided gold particles recovered.

ADVANTAGES AND DISADVANTAGES OF AMALGAMATION

Due to the fact that in most ores a considerable portion of the gold is attached to quartz or associated with base metal sulphides, even after fine grinding, the loss of gold by the amalgamation method is considerable as compared to cyanidation, and, for this reason, amalgamation is usually followed by either cyanidation, tables, or flotation in order to increase the recovery of gold in these forms. In general, cyanide treatment will recover part of the gold attached to quartz or associated with iron sulphide and tables or flotation will also partly recover the gold in these two forms. While it is a fact that amalgamation usually makes a low recovery as compared to other methods, it has the advantage that the product produced is gold bullion, readily and cheaply marketed. The amalgamation process also has the advantage of being easier to operate as compared to cyanidation and flotation and, while, as in other things, experience is necessary for efficient work, it takes less time, on the whole, for a man to become proficient in the practice of amalgamation than either cyanidation or flotation.

AMALGAMATING PRACTICE

Grinding of ores prior to amalgamating is usually done by gravity stamps, ball mills, Huntington mills, or arrastras.

GRAVITY STAMPS

Figure 1 is a sectional view of a stamp battery, from Richards' "Textbook of Ore Dressing." Briefly, it consists of a battery box, bolted firmly to a heavy concrete foundation and fed from an ore bin by hand feeding or a mechanical ore feeder. The ore, previously crushed to between three-inch and one-inch sizes, is crushed further within the battery box by means of stamps raised by two-way cams and dropped by gravity through distances of six to ten inches at the rate of ninety to one hundred drops per minute per stamp. A screen placed in front of the battery box confines the ore within the box until crushed to the desired size. Water, added at the rate of four to seven tons per ton of ore treated, is fed with the ore. Mercury is usually added to the battery with the ore by means of a wooden spoon, the amount varying from one to two ounces per ounce of gold amalgamated. The addition of mercury to the stamp battery is of advantage to the small operator in that it catches part of the gold in coarse condition within the battery. Plates within the battery utilize the impinging action caused by the splash of the ore pulp when the stamps drop which throws gold particles against the plates, a condition favorable, as previously mentioned, to amalgamating fine gold. The disadvantages of inside plates include the difficulty of keeping them in good condition, since they cannot be examined readily, and the scouring action of the ore pulp.

The rate of feeding ore to the stamps is controlled by the sound or feel of the blow struck by the stamps. A soft, mushy blow indicates over-feeding while a metallic blow with excessive rebound of stamps indicates under-feeding. Over-feeding of stamps should be especially avoided since very little crushing takes place within an over-fed battery and it takes considerable time to clear out a choked battery after feeding of ore has been stopped.

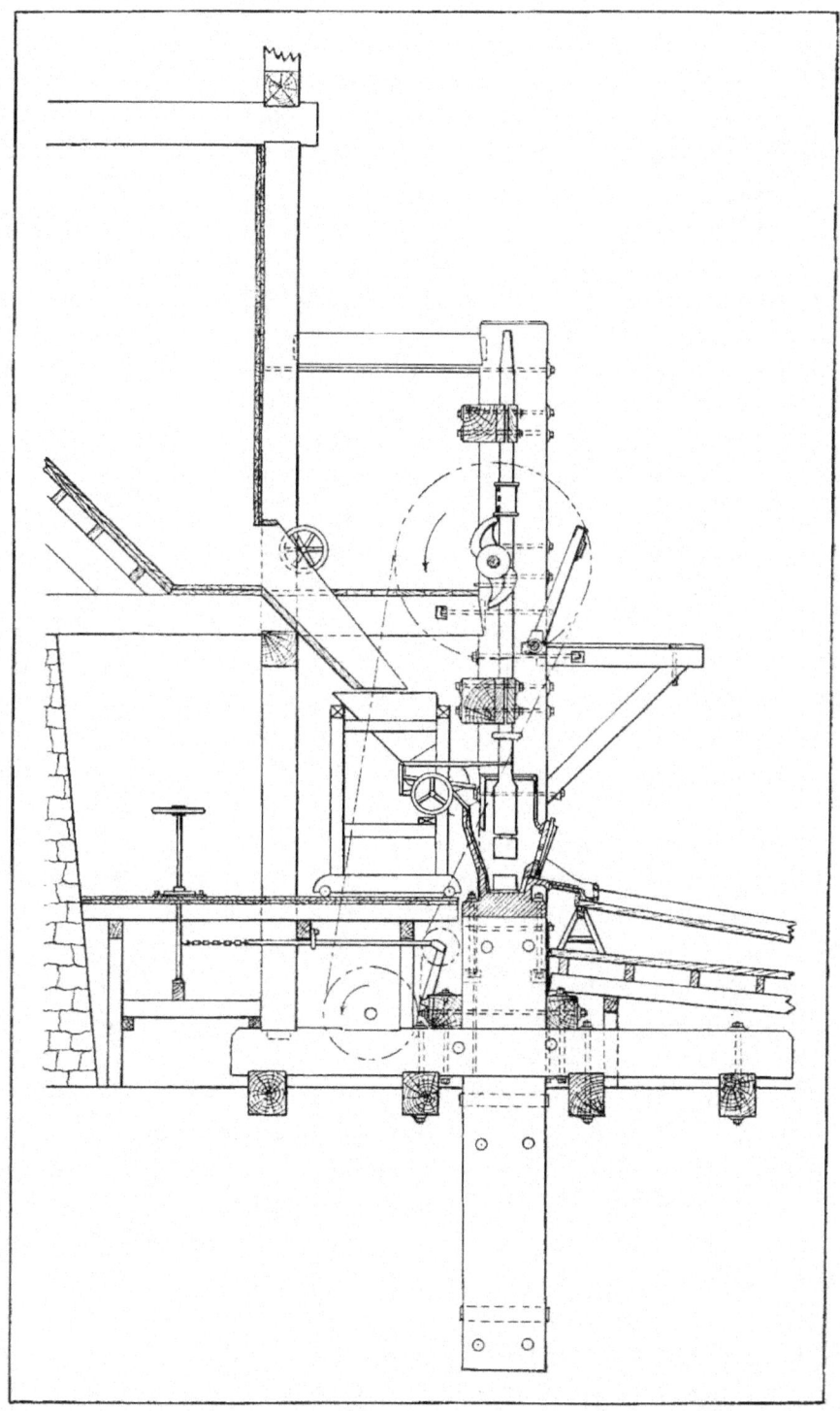

Fig. 1. Stamp Battery. Taken from Richards' "Textbook of Ore Dressing."
McGraw-Hill Book. Co., New York City.

Occasionally a shoe becomes loosened and drops off the boss. The whole battery must then be shut down in order to refasten or replace the shoe. Soft pine wedges, which swell after wetting in service, are usually used for holding the shank of the shoe to the boss.

In order to avoid excessive whirling of the stamp, which is caused by friction with the cam, regular greasing of the cam is necessary, and, due to the harmful effects of grease upon mercury, this greasing must be done carefully and any excess of grease avoided. A rag rubbed on hard or medium-hard grease and then held against the contact surfaces of the cam, as it revolves in operation, will place on the cam a thin film of grease which is sufficient to prevent excessive whirling of the stamp.

The height of drop of the stamps is varied within certain limits by raising and lowering the tappet on the stamp stem. The number of drops of a given stamp per minute is limited by the height of drop and by the time consumed per drop. If more than 100 drops per minute are attempted, there is danger of the tappet striking the arm of the cam and breaking it during the descent of the stamp.

When the ore pulp passes through the battery screen, it drops to a copper plate placed upon a slight slope. This copper plate is coated either with mercury or silver amalgam and a second catch of gold is effected on this plate.

Upon leaving the plate, the ore pulp passes through a trap for the recovery of mercury and amalgam which has been scoured off the plate by the pulp stream. The ore pulp is then, at times, as previously stated, treated by either tables, cyanidation, or flotation in order to recover fine gold or gold attached to quartz or combined with base metal sulphides.

At certain time intervals, twelve to twenty-four hours, amalgam is removed from the plate, especially from the upper portion of the plate; it is first softened by adding more mercury and then scraped together with a rubber squeegee and removed. The plate is then dressed by rubbing with a rag and any stains removed with a small amount of cyanide. During the operation of the stamps, stains are also removed from the plate, when necessary, by means of cyanide as already indicated. A convenient method of adding mercury to plates is by means of a mercury "shaker" which is made by partially filling a four-ounce, wide-mouthed bottle with mercury and tying a small piece of chamois or fine canvas over the mouth.

Although many materials have been used for amalgamating plates, soft, annealed, rolled-sheet copper, not less than 1/16 inch in thickness, is the most common plate material used. Richards and Locke [1] describe the Louis method of preparing new copper plates for service as follows: "Fine sand (sea sand if obtainable) is sprinkled on the plate, well moistened, and rubbed in with a block of wood until every portion of oxide is removed and the plate has a uniform red surface, care being taken at the same time not to scratch it. The sand is then washed off, and the plate dried and polished with fine emery paper folded over a block of wood. A perfectly clean, dry surface is thus produced. A mixture is then made of about ten parts sand to one part of coarsely pounded sal ammoniac; this mixture is dampened with water and clean, pure mercury is sprinkled into it by squeezing through canvas. The mixture is then rubbed over the plate with a piece of canvas or blanket when amalgamation will at once begin; more mercury must be sprinkled on the plate from time to time, and the rubbing continued until a uniformly bright, silvery surface is obtained. Each square foot of copper will require about 1/2 ounce of mercury. The plate is next washed well with water and kept until the following day. It will then probably be found that the plate is dulled and covered with a coating of a greenish-gray substance. Usually the plate is brightened with a dilute solution of cyanide, together with a little mercury."

The Louis method, as described, prepares new plates by removing all grease and copper oxide surface coatings and then puts a thin coating of copper amalgam on the plate. It has sometimes been found advisable to use a coating of silver amalgam rather than copper amalgam for new plates since silver amalgam will insure a better recovery of gold than copper amalgam when the latter is used on new plates. A convenient method of preparing silver amalgam follows.

Use sufficient silver to allow about 1/4 ounce of silver per square foot of plate surface. Dissolve the silver in nitric acid (1.2 specific gravity). Remove the excess acid by evaporation, but do not bake the residue. Re-dissolve the silver nitrate in distilled water, using a small amount of nitric acid if found necessary to obtain complete solution of the silver. Dilute the solution to between 1,000 cc. and

---

1. Richards, R. H. and Locke, C. E. Texbook of the Ore Dressing, p. 57, McGraw-Hill Book Co., New York City.

2,000 cc. (1 to 2 quarts) and place strips of sheet copper in the solution to precipitate the silver in a very finely divided form. After 12 to 24 hours remove the copper strips with attached silver to a separate glass container and dilute with one quart of distilled water. Heat almost to boiling, remove the copper strips and allow the fine silver to settle. Decant off the water and repeat the washing five to eight times to remove all copper. Finally, dry the silver and add mercury to it until the resulting amalgam can be easily molded in the hand but is not sufficiently liquid so that liquid mercury is oozing out of the mass (about three parts of mercury to one part of silver is required). After cleaning the plate by scouring as described by the Louis method, the silver amalgam is painted on the plate by rubbing the ball of amalgam back and forth sidewise (but not lengthwise) on the plate and finally smoothing the amalgam to a more or less uniform thickness with a clean brush or whisk broom.

At intervals varying from a week to a month, the stamp battery is shut down and, after removing the screen, shoes, and dies, the battery is cleaned out. At these clean-up periods, the amalgamating plate is scraped with putty knives and polished. After removing the accumulated amalgam and residue in the battery, repairs are made and the battery reassembled for another run.

The amalgam is usually recovered from the battery residue by panning, and, together with the amalgam recovered from the plate, is first cleaned by washing and then squeezed through chamois or other material. The mercury that passes through this filter is nearly free of gold and may be used again. The solid amalgam which is retained by the filter is ready for separation of the gold and mercury by retorting.

RETORTING

A retort for small-scale work is shown in Figure 2 and consists of an iron crucible equipped with a tight fitting cover held in place by a clamp. A hole is bored into the cover and a bent iron pipe for removing the mercury fumes is fitted into this hole. The discharge end of this bent pipe is surrounded by a water jacket through which cold water circulates and causes the mercury fumes to condense to liquid mercury. Upon heating the retort, the mercury contained in the gold amalgam vaporizes and enters the iron pipe where it is condensed to liquid and finally drops into a container partly filled with water, provided for the purpose. The gold remains in the retort.

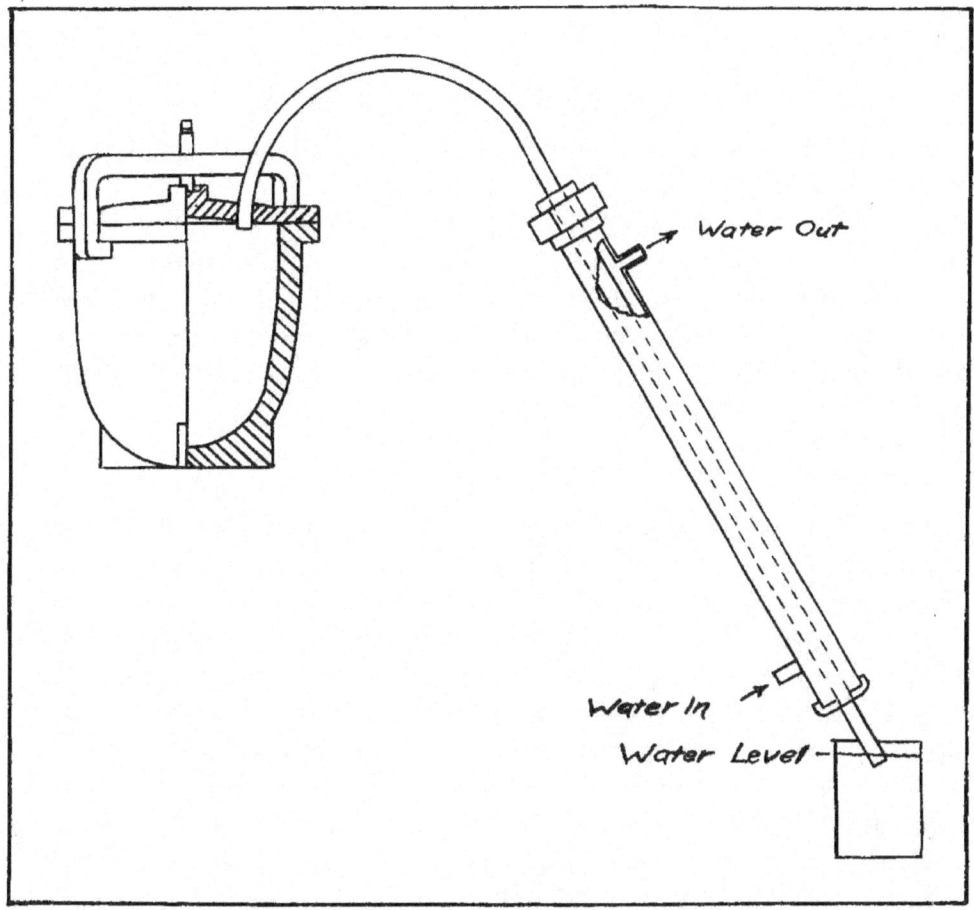

Fig. 2. A retort.

The inside of the retort is usually coated with a very thin chalk or clay emulsion to prevent the gold bullion from sticking to the retort. A double thickness of newspaper also serves the same purpose. The cover, after being fastened in place by a clamp, is luted with chalk or clay emulsion. The retort should not be filled more than two-thirds to three-fourths full and the discharge pipe should barely dip beneath the surface of water so that all the fumes will condense and, when the retort cools, although water may be drawn into the pipe, water will not reach the crucible and cause it to explode.

It should be remembered that the fumes of mercury are poisonous, and, for this reason, the retorting operation is best performed in the open, using a wood fire. The retort crucible should be allowed to cool before opening. If the operation is conducted indoors, special precautions must be taken to prevent the mercury fumes from concentrating in the atmosphere.

MELTING AND REFINING OF RETORT BULLION

The residue remaining in the retort after the removal of the mercury is usually placed in a graphite crucible and melted with fluxes. The fluxes commonly used for this pupose are niter, silica, sodium carbonate, and borax. After thorough melting and stirring, the contents of the crucible are usually poured into an iron mold previously coated with lubricating oil. The slag contains some gold and should be saved for re-treatment or sold to a smelter after a sufficient amount has accumulated.

BALL MILLS

Ball mills have replaced gravity stamps to a considerable degree during recent years in grinding ores prior to amalgamation in large-capacity plants especially when it is desired to follow amalgamation with very fine grinding for cyanidation. In small-capacity plants, however, the gravity stamp mill retains favor, especially. when treating hard ore, because it is easier to regulate small-capacity gravity stamp mill units to the desired capacity than is possible when ball mills are used. Gravity stamps are also convenient amalgamators whereas certain difficulties, including the flouring of mercury when placed in the ball mill and the segregation of coarse gold in the ball mill, make the operation of amalgamation somewhat more difficult in ball mills than in stamps. Ball mills, however, can be used to advantage if proper precautions are taken for amalgamation and the choice between stamps and ball mills will depend largely upon existing conditions.

HUNTINGTON MILLS

Huntington mills have been used in the past for crushing ores prior to amalgamation. A mill of this type consists of rollers with vertical axles, which operate against a circular die or ring in the rim of a pan. Crushing is facilitated by centrifugal force which drives the rollers against the ring die. The shipping weight of this mill per ton of daily capacity is less than for gravity stamps and this fact is the chief point in favor of such a mill. The rollers, however, do not usually wear smoothly, and the rollers then pound against the die. Although the machine is a good amalgamator, it is not in favor at present owing to the high operating and maintenance charges as compared to stamps and ball mills.

ARRASTRAS

The arrastra is a device for doing the work of a stamp mill, but most of the materials used for its construction are available in the vicinity of the mine. A drawing of an arrastra is presented in Figure 3. Richards and Locke, in their "Textbook of Ore Dressing," describe arrastras in general as follows: "This mill consists of a circular pavement from six to twenty feet in diameter, with a retaining wall around it and a step in the center. Upon the step stands a vertical, revolving spindle or shaft, and from the spindle extend horizontal arms to which large boulders, called drag-stones, are attached by chains. The boulders are dragged around the circle by arms and crush the ore by a true grinding action.

"The arms number from two to eight, usually four. The drag-stones vary from two to twelve, commonly four; they weigh from 80 to 2,000 pounds, averaging about 300 pounds. Holes are drilled in the stones, plugged with dry wood, and the eye rings are driven into these plugs. They are placed so that the stone shall slide on its largest surface and forward of the center of gravity so that the front edge of the stone may be lifted sufficiently to slide over the coarsest of the ore during the early stage of grinding.

"To prevent leakage of quicksilver, the pavement is built upon a clay or concrete foundation which is always wider than the pavement. The latter is almost one foot thick, of granite, basalt, or flinty quartz, a rough-grained rock being preferred. The joints are filled with fine tailings, or, better, with cement. The retaining wall, two to four feet high, is made of stones laid in cement, of wooden staves bound with iron hoops, or is merely a clay bank. It has a gate or a series of plug holes for discharging the pulp and, sometimes, screen discharges for continuous work.

"The speed is four to eighteen revolutions per minute, usually ten to fourteen for power arrastras. Small arrastras are driven by a horse or mule attached to an extension of one of the arms, the animal walking around the circle. Large arrastras are driven by a horizontal water wheel, suspended from cross arms separate from the dragging arms and extending outside the retaining wall, or they

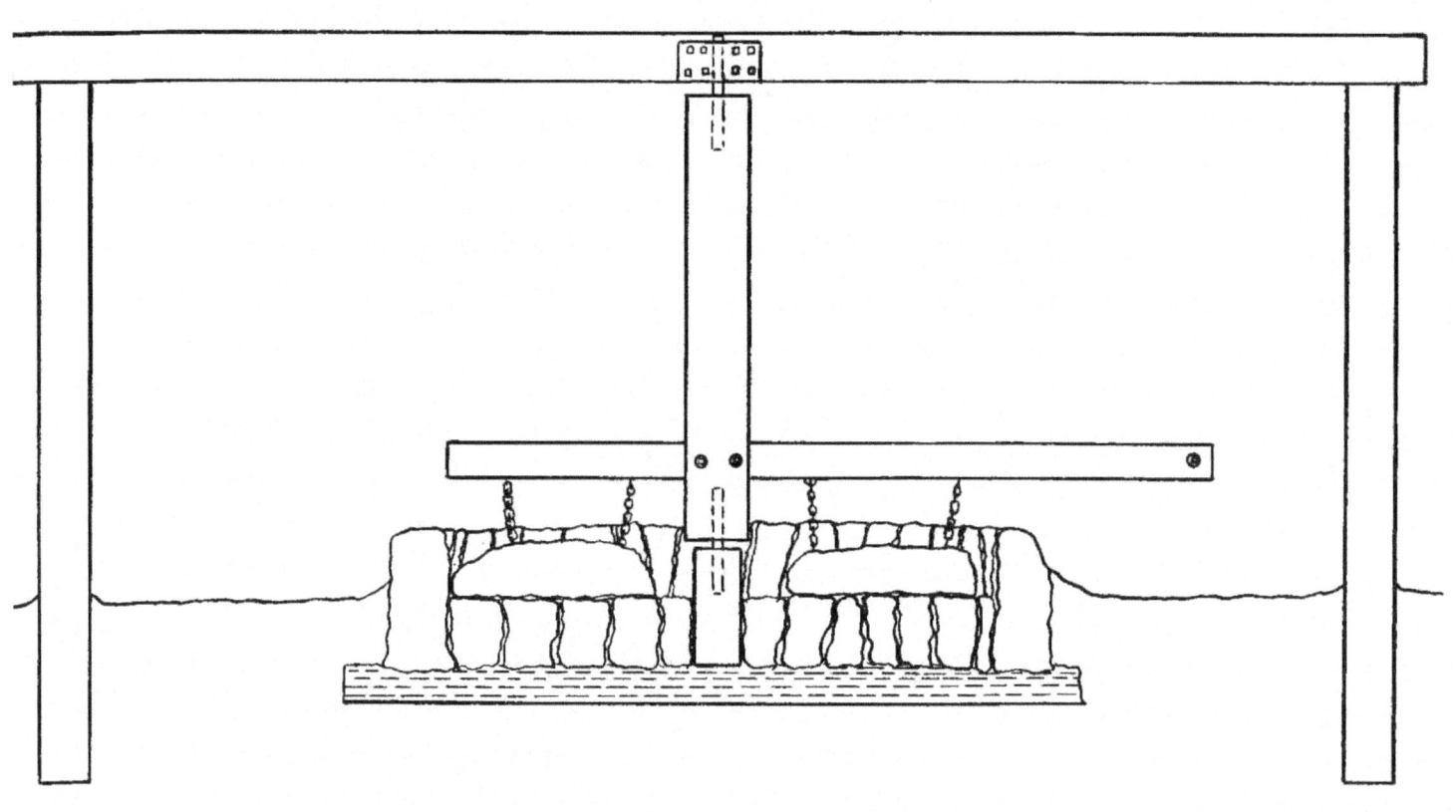

Fig. 3. Arrastra.

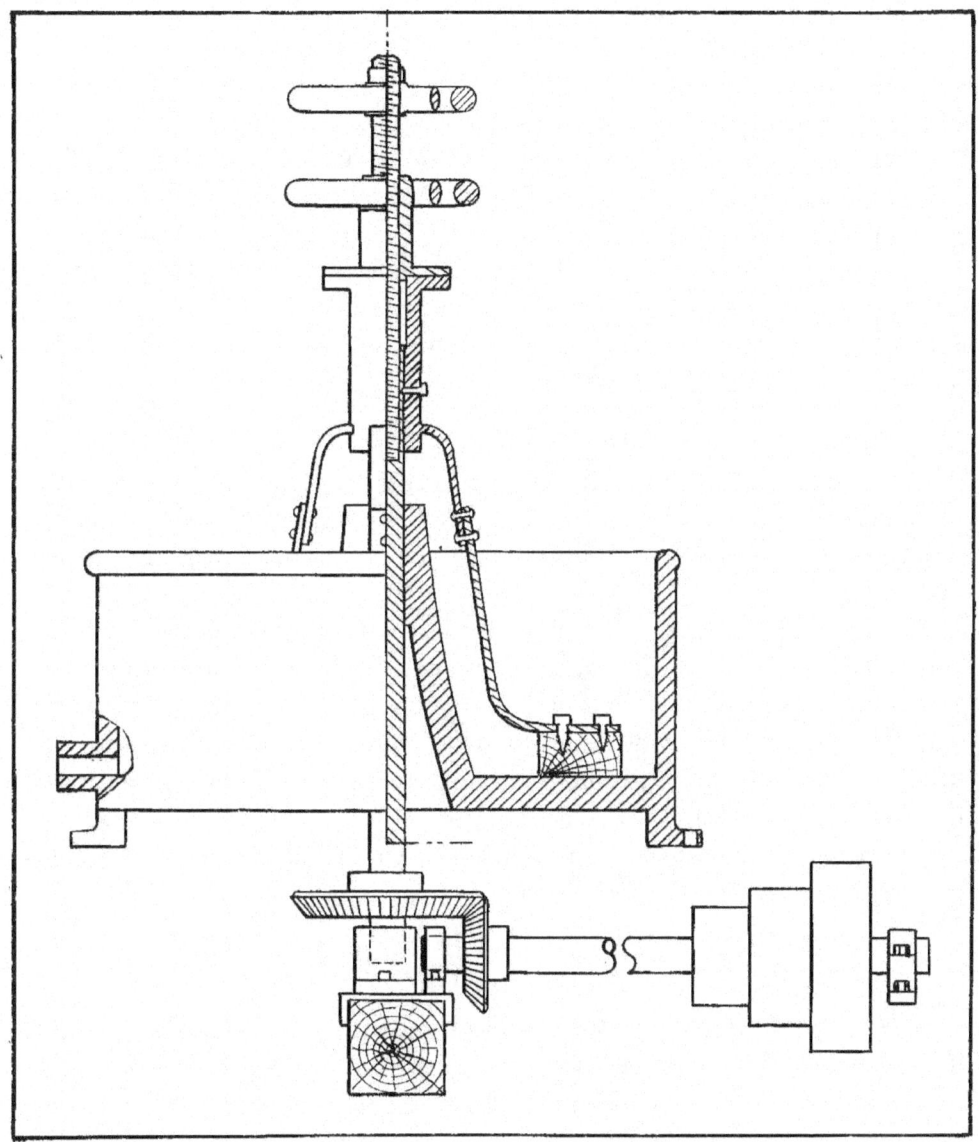

Fig. 4. Amalgamating pan.

are driven by a shaft with beveled gears. One long shaft may in this way connect several arrastras with a single driving engine.

"It is used as a fine grinder and amalgamator with both gold and silver ores, and is fed with material seldom above 3/4 inch in diameter, often much below. It is used where cheapness, both of installation and of running, is essential and, at the same time, small capacity is not objectionable, for example, in regions remote from supplies. It is often used for re-treating tailings of gold mills, chiefly by lessees."

Arrastras are usually operated intermittently. A batch of ore is added with sufficient water to make a thick pulp when the ore has been ground. The mercury is usually added at the last stage of the grinding operation. After sufficient time has elapsed for amalgamation of the gold, the contents of the arrastra, including the amalgam, are removed and the amalgam is recovered by panning or in rockers. If sufficient water is available, the charge may be removed from the arrastra by sluicing and passed through traps for the recovery of the amalgam.

AMALGAMATING PANS

The amalgamating pan shown in Figure 4 is a device for cleaning up the residue of a stamp battery or for cleaning up black sands from placer work. A rotating barrel is often used for the same pupose.

The grinding pan comprises a cylindrical container into which the material to be amalgamated is placed with water and mercury. The pulp is mixed and stirred by means of revolving framework called a muller. The pulp should be sufficiently thick to maintain finely divided mercury globules in suspension but not thick enough to prevent movement of the globules in the pulp. After a few hours of stirring, the mixture is removed and the amalgam separated by pans or rockers.

COST OF RECOVERING GOLD BY AMALGAMATION METHODS

It is difficult to give costs of amalgamating gold ores that would be useful to small operators since operating milling costs in small plants depend to a large extent upon the character of the ore treated and operating conditions. Operating conditions vary greatly in such factors as per cent of operating time obtained by the mill, cost of water, power, and labor, and accessibility of the mill to supplies.

For the purpose of indicating cost of amalgamating gold ores on a small-capacity basis, the following costs are given for a 25-ton capacity amalgamation mill in Canada. The figures have been taken from United States Bureau of Mines I. C. 6433, Amalgamation Practice at Porcupine United Gold Mines, Ltd., Timmins, Ont., by R. A. Vary, March, 1931.

OPERATING COST PER TON OF ORE TREATED

	Labor	Power	Supplies	Total
Crushing	$0.214	$0.135	$0.035	$0.384
Grinding	0.429	0.180	0.162	0.771
Classifying, screening, conveying and refining....	0.429	0.090	0.107	0.626
Miscellaneous	....	....	0.070	0.070
Totals	$1.072	$0.405	$0.374	$1.851

As previously mentioned, these figures are to be taken only as an indication of cost per ton in a well-operated 25-ton capacity plant. For a mill less efficiently operated, less accessible to supplies, or treating a more refractory ore, the cost would be greater. It should also be noted that the costs given are operating costs only and do not take into consideration writing off the first cost of building and equipping the mill. For small ore bodies, this latter cost is considerable and is at times the controlling factor between profit and loss on the undertaking as a whole.

Richards and Locke give combined crushing, stamping, amalgamating, and concentrating operating costs from data collected by the General Engineering Company of Salt Lake City for a period comparable to present unit power, labor, and supply costs. This tabulation is given to show the effect of plant capacity upon operating costs and indicates the disadvantage to which the small opertor is put in this respect.

Capacity, tons per 24 hours	Location	Description of treatment	Cost per ton
30	Calif.	Crushers, gravity stamps and amalgamation	$1.92
50	Calif.	Crushers, gravity stamps, amalgamation, gravity, concentration	1.07
300	Calif.	Crushers, gravity stamps, amalgamation	0.22
1000	Alaska	Crushers, gravity stamps, gravity concentration, and amalgamation	0.24

CYANIDATION

The cyanide method of recovering gold from ores is based upon the principle that dilute solutions of sodium or potassium cyanide (sodium cyanide is commonly used) will dissolve gold from crushed or ground ores under certain conditions. After the gold has been dissolved, the solution is separated from the ore residue and the gold recovered from it by means of metallic zinc shavings or dust.

The cyanide method, in contrast to amalgamation, flotation, and gravity concentration methods, is distinctly a chemical method and cannot be used to advantage without chemical control of cyanide quantities, alkalinity, and assays of solutions and residues.

There are two general methods of applying the cyanide process to gold ores, namely: the percolation method and the agitation method. The size to which the ore must be crushed or ground prior to treatment in order to obtain satisfactory extraction of the gold within reasonable time periods largely governs the amount of slimes produced and is usually, therefore, the determining factor of deciding which of these two methods is to be used in small-scale operations.

In the percolation method, as applied to the entire ore, the material is crushed to between 1/4 inch and about 10-mesh and placed in vats, usually made of redwood and equipped with false bottoms of cocoa matting and canvas, through which the solution, but not the ore, may penetrate. The cyanide solution, ranging in strength from 0.25 to 0.05 percent, is added to the tank and slowly allowed to percolate through the ore charge. After sufficient time has been allowed for satisfactory solution of the gold, the pregnant gold solution is clarified and, in small-plant operations, is usually passed through boxes filled with zinc shavings to precipitate the gold. After the tank containing the ore residue has drained, barren solutions (solutions after passing through zinc boxes) are used as washes and, finally, fresh water is added to remove as much as possible of the dissolved gold remaining in the residue. Lime is usually added to the crushed ore before applying the cyanide solution in order to neutralize any acidity in the ore and thereby save the expensive cyanide salt which is consumed by acid in the absence of lime.

In the agitation method, the ore is finely ground, usually in a cyanide solution, and, after grinding, the thick ore pulp is agitated in tanks by means of mechanical stirrers or compressed air for a period of time sufficient to obtain satisfactory solution of the gold. The gold-bearing solution is then separated from the solid ore residue by repeatedly diluting and decanting the liquid or by filtering. The gold is then precipitated from the clarified solution by metallic zinc.

The first method described, the percolation method, has two definite advantages over the second or agitation method, especially for the small operator, namely, lower first cost of plant and lower operating cost. The lower first cost results from the coarser crushing used in this method and the absence of mechanical agitators, filtering equipment, and the necessary power plant equipment to operate them. The lower operating cost is due to coarser crushing, elimination of power for agitation and filtration, and the simplicity of the process as compared to the agitation method.

The disadvantages of the percolation method as compared to the agitation method are lower extraction of the gold and longer periods of time usually required to extract the gold.

The percolation method should be thoroughly tested before installing, not only for such information as percent recovery of gold, consumption of cyanide, time required for satisfactory gold recovery at size of crushing decided upon, but, also, for satisfactory percolation rate of solutions through the ore charge when using the same depth of ore charge that is planned to use in actual operation. This latter precaution is very important since many small-size plants have failed to operate successfully due to the fact that these preliminary tests were not made. It should be noted that, if a solution percolates satisfactorily through a column of crushed ore six to twelve inches high in a laboratory percolator, it does not necessarily follow that solutions will percolate through ore columns three to five feet high at a satisfactory rate in actual operations. Resistance to the percolation of solutions depends upon the depth of ore charges, the size of crushing, and the amount of slimes contained in the charge. Before deciding upon the installation of percolation equipment, a test should be made in which the depth of ore charge, the size to which the ore is crushed, and the amount of slimes are the same, as nearly as possible, as they will be in actual operations.

The agitation method, as indicated, possesses the advantages of higher recovery and shorter time of treatment as compared to the percolation method, owing to finer grinding used. Its high initial and operating costs are, however, serious obstacles to its installation for small-plant operations. In large-scale cyanide operations where maximum recovery is the governing factor, the ore must be finely ground, and the agitation method is then usually required since cyanide solutions will not percolate, as a rule, through finely ground ore beds. In small-scale operations, the first cost of the plant is at times the governing factor, and, therefore, in small-scale operations, it is sometimes more profitable to treat the coarsely crushed ore by percolation and accept the lower recovery than to attempt to extract more of the gold by a plant too expensive for the size of the ore body.

It is not possible to determine, without testing, whether an ore is or is not amendable to cyanide treatment. In general, ores containing finely divided metallic gold are more amendable to straight cyanidation than ores containing coarse gold since the time required to dissolve a given quantity of fine gold is less than that required to dissolve the same amount of coarse gold. Amalgamation can, however, be used prior to cyanidation and the coarse, free gold recovered at low cost by amalgamation, leaving the fine, included, and attached gold to be partly saved by cyanidation.

Besides the size of gold particles in ores, other characteristics of ores influence the amendability of ores for cyanide treatment. Among the most important of these characteristics, the following are briefly noted.

Gold, attached to or included in quartz or iron pyrite, is dissolved by cyanide providing the solution reaches the gold and sufficient time is available. Gold lost by the amalgamation method, due to its presence in these forms, is, therefore, partly recovered by cyanidation, and it is for this reason, chiefly, that recoveries of gold by cyanidation are usually higher than by amalgamation.

Sodium cyanide is a relatively expensive chemical reagent costing, at present, from twenty-five to thirty cents per pound. The consumption of this reagent is, therefore, not only important as affecting cost of treatment, but is, at times, prohibitive. Cyanide is lost mainly in two ways: First, chemically, by reacting with various impurities contained in the ore, and, second, mechanically, due to imperfect washing of ore residues. Acids derived from organic matter and from oxidation of pyrite consume cyanide, but this loss is largely prevented by adding a cheaper alkali, lime or soda ash, to the crushed or ground ore prior to the addition of cyanide. The lime or soda ash reacts with the acid and thereby saves the cyanide.

Copper minerals, especially oxidized copper minerals, and iron compounds (ferrous) are the most common cyanide consuming impurities that lead to high and, at times, prohibitive cyanide consumptions. Although examples can be given of large plants operating successfully with these impurities present in the ore, it requires considerable skill and high plant investment to use the cyanide method profitably when appreciable amounts of these impurities are present in the ore.

An important condition that is encountered in cyanide operation is the fouling of solutions. This condition is caused by the accumulation of impurities in the solutions as they are used over and over again. The solutions may also deteriorate if certain impurities in them consume oxygen since gold will not be dissolved unless oxygen is present in the solutions. The chief effect of the fouling of solutions is either to reduce the amount of gold dissolved from the ore or to increase the time necessary for dissolving the gold. Both of these effects are serious in plant operation, and advance information as to the probable results of fouling can only be obtained by small-scale, continuous laboratory tests whereby the laboratory solutions are allowed to foul.

The gold precipitate produced by contact of the gold-bearing solutions with metallic zinc, although impure, can be marketed directly to a smelting plant. It is usually better, however, to remove most of the zinc that is mixed with the precipitate and then to dry and melt the material with proper fluxes in order to produce gold bullion bars which can be sold directly to the United States Mint. A large amount of the zinc that is mixed with the gold precipitate may be removed by washing the precipitate on screens if zinc shavings were used for the precipitation operation. The screen treatment may be followed by sulphuric acid to remove additional zinc or the sulphuric acid treatment may be used directly if zinc dust has been used to precipitate the gold. The washed, cleaned precipitate is then nearly dried and melted with sodium carbonate, silica, and borax fluxes.

As already indicated, the cost of cyanidation depends upon the method selected and upon the character of the ore treated. Straight cyanidation by the percolation method is a cheap process when applied to a clean ore, but few ores can be treated by percolation alone since fine crushing of the ore is usually necessary for satisfactory recovery and this fine crushing in turn produces a product which contains too much fine material to be handled by this method.

The percolation process has often been used on the coarser sand portion of an ore after comparatively fine grinding while the fine slimes are treated by the agitation cyanide method or by flotation. The Homestake Mining Company, in 1929, crushed, ground, amalgamated, and cyanided the sand by percolation and the slimes by agitation for a total operating cost of $0.503[1] per ton of ore. The total cost was distributed as follows: $0.038 for coarse crushing, $0.276 for grinding and amalgamation and $0.189 for cyanidation. The efficient Homestake operation, conducted on a basis of about 4,000 tons per day, cannot of course be duplicated in a small plant as far as cost is concerned. Under favorable small-plant operating conditions, the operating cost for the percolation method may be as low as $0.50 to $0.75 per ton, and, under less favorable conditions, it might run up to from $2.00 to $3.00 per ton or higher. The operating cost for the slime agitation process, when operated on a small-capacity basis, will range from $3.00 to $6.00 per ton of ore treated.

GRAVITY CONCENTRATION

Although the gravity concentration method is often used without the aid of accessory methods in recovering gold from placer deposits, it is not often so used in the treatment of ores derived from lode deposits. In the treatment of the latter material, gravity concentration is more often used as an accessory method to recover part of the gold which escapes amalgamation, cyanidation, or flotation. Wet concentrating tables and vanners are commonly used for this purpose and many types are available. Some of them have stationary, sloping surfaces covered with canvas, blankets, cement, or riffles. Others are provided with movable decks which are either given a rapid forward and backward motion or, in the case of vanners, a more gentle sidewise or lengthwise, swaying motion.

Tables take advantage of the facts that gold particles are not carried along by a current of water as rapidly as are the other constituents of the ore, and they settle more rapidly than other constituents of the ore in a liquid ore pulp and, therefore, find their way to the bottom of the pulp bed. While settling will take place without motion of the table deck, it is usually desirable to move

1. Clark, A. J., Milling methods and costs at the Homestake Mine, Lead, S. Dak., I. C. 6408, U. S. Bureau of Mines, Feb. 1931.

the deck since the separation is then hastened and the treatment is continuous. In the case of very fine gold particles, however, motion of the table interferes with the recovery of gold, and, for this reason, stationary canvas- or blanket-covered tables will effect a higher recovery of fine gold than movable tables of the Wilfley or Deister types. The capacities of stationary tables are, however, much less than movable tables and they operate intermittently, whereas the movable decks operate continuously, as stated. Intermittent operation means that the table is fed for a certain period during which time the pores of the canvas or blanket become filled with the minerals desired to be recovered; the feed is then stopped and the values removed after which the operation is repeated.

The moving power of a current of water is a very important feature of any wet concentrating table. The water which flows down the gentle slope of the table or vanner deck removes the waste material while the heavier minerals are retained by riffles, canvas, or blankets.

The usual cost of small-scale table and vanner treatment, not including the cost of crushing, grinding, screening, or other accessory operations, ranges from $0.05 to $0.25 per ton and depends largely on capacity handled and cost of water.

FLOTATION

The use of flotation methods in the treatment of gold ores is a comparatively recent development and offers considerable promise to the small operators in that the first cost of the plant and the operating cost should be well under the figures required for a fine grinding agitation type of cyanide plant. Although the recovery of gold made by a flotation plant will not usually be as high as in the competing fine grinding cyanide plant and although the flotation concentrate is at times shipped to a smelting plant whereas gold bullion with its cheap marketable feature is obtained in a cyanide plant, nevertheless the lower first and operating costs of a flotation plant will, it is believed, in many instances, more than offset the advantages of cyanidation. Again, it is at times possible to cyanide the flotation concentrate produced and, when this is the case, one of the advantages of cyanidation is eliminated.

In the flotation process, the ground ore, mixed with water, is brought in contact with bubbles of air. The gold particles, possessing a bright, metallic surface, tend to adhere to the air bubbles whereas the earthy, waste constituents of the ore, having a dull, non-metallic luster, have less tendency to cling to air bubbles. The reagents usually used for the floating of gold consist of lime, a collector, and a frother. The amount of lime needed varies for different ores and can be determined only by testing. The collector reagent smears the surfaces of the gold particles and thereby increases their adhesion to the air bubbles. The results of experimental work done by the writer show that amyl xanthate and sodium aerofloat are satisfactory collectors in the treatment of gold ores. These reagents are used in small amounts, from 0.03 to 0.20 pound per ton of ore treated and recent work by others indicates that smaller quantities at times produce better results. The frother is added to prevent the bubbles from breaking and, for gold work, steam-distilled pine oil, added at the rate of 0.05 to 0.10 pound per ton, has been found satisfactory.

There are many types of machines on the market for the treatment of gold ores and the chief difference in the principles of these machines is the manner of supplying the air bubbles. The type usually used for the treatment of gold ores is the mechanical agitation machine with or without sub-aeration. Air is supplied to the straight mechanical agitation type machine by beating it into the liquid pulp by the use of rapidly revolving arms or paddles. In sub-aeration mechanical machines, additional air is usually provided by a blower, at a low pressure. The ore pulp, with reagents, is thoroughly mixed with the air in the agitation compartment of the machine and then discharged into a compartment for settling. The bubbles of air, with gold particles attached, form a froth which rises to the surface of the ore pulp from which position it is removed.

The second important type of machine is the pneumatic which receives all the air used from a low pressure blower. Some pneumatic machines are provided with a canvas mat at the bottom through which the air from the blower must pass before mixing with the ore pulp. The pneumatic machines that are not provided with mats are called either "matless" or air-lift type machines.

Results of comparative tests of the mechanical and pneumatic type machines, when operating with gold ores, have not been published, to the writer's knowledge. There is, however, a feeling among mill men that the mechanical type of cell is preferable for the treatment of gold ores, and tests made by the writer show this type of cell is satisfactory for gold ores. As a rule, the power cost for operating the mechanical type cell is higher than that for the pneumatic cell, but, due to more efficient mixing of reagents, the mechanical cell usually is more economical in cost of reagents and, in some instances, possibly, is superior from the standpoint of recovery.

With small plants operating on gold ores, it is believed that tables operated as scavengers, following flotation, will be found valuable. The operation of a table, as described, does not greatly increase the operating cost and it will serve as a pilot machine for observation of flotation results and will, in many instances, recover an appreciable percentage of the total gold.

In the flotation treatment of ores containing some coarse gold, amalgamation may be useful. From the results of tests made by the writer, amalgamation, if carried out, should be used before adding the flotation reagents to the ore as these reagents have been found to have a harmful effect upon the mercury.

In large flotation plants, the operating cost, including crushing, grinding, and all accessary operations, varies from $0.18 to $1.25 per ton of ore treated. With small plants this cost will vary from $1.00 to $2.50 per ton of ore treated, and the maximum figure given may be exceeded if the percent of operating time is not maintained at a satisfactory figure.

In order to show the distribution of operating costs and metallurgical results of a 150-ton capacity mill, operating on gold ore by flotation methods, the following figures, recently published by the U. S. Bureau of Mines, are given for the Spring Hill Concentrator located near Helena, Montana.

SUMMARY OF MILLING COSTS AT THE SPRING HILL CONCENTRATOR, JULY 1, 1929 TO MARCH 3, 1930.

Cost per ton of ore milled[1]

	Labor	Super-vision	Supplies	Reagents	Power	Total
Crushing and conveying	$0.062	$0.001	$0.050		$0.049	$0.162
Grinding	.100	.001	.072		.117	.290
Flotation	.108	.001	.055	$0.063	.031	.258
Disposal of tailings	.048	.001	.003		....	.052
Water	....	.001	.001		0.010	.012
Repairs and maintenance	.102	.001	.130		....	.233
Total	$0.420	$0.006	$0.311	$0.063	$0.207[2]	$1.007

[1] Grant, L. A., Milling Methods and Costs at the Spring Hill Concentrator of the Montana Mines Corporation, Helena, Mont., I.C. 6411, U. S. Bureau of Mines, March, 1931.

[2] Power costs $0.008 per kilowatt hour.

METALLURGICAL DATA AT SPRING HILL CONCENTRATOR JULY 1, 1929 TO MARCH 31, 1930.

Heads assay...................$6.46
Tailings assay...................$1.22
Total ore concentrated, tons...................40,930
Days operated...................270
Operating time per day, hours...................24
Operating time, per cent...................98
Average ore concentrated per 24 hours, tons...................153
Recovery of gold, per cent...................82
Concentration ratio...................12 to 1
New water consumption per ton of ore concentrated, gallons...................104
Ball consumption per ton of ore concentrated, pounds...................1.30
Reagent consumption, pounds per ton of ore milled
  Pine oil...................0.075
  Potassium ethyl xanthate...................0.109
  Aerofloat...................0.103

MAPS OF ARIZONA

The Arizona Bureau of Mines now has available for distribution three different maps of the State, as follows:

1. Base map of Arizona in two sheets on a scale of about eight miles to the inch. This map is strictly geographic, with the positions of all towns, railroads, rivers, surveyed lands, national forests, national parks and monuments, etc., indicated in black, and the location of mountains and other topographic features shown in brown. It also indicates where the various mining districts are situated, and is accompanied by a complete index. It was issued in 1919 and is sold unmounted, for 35c., or mounted on cloth with rollers at top and bottom for $2.50.

2. A topographic map of Arizona in one sheet, on the same scale as the base map. It shows 100-meter contours, and there is a meter-foot conversion table on the map. It was issued in 1923, and is sold, unmounted, for 50c, or mounted on cloth with rollers at top and bottom for $2.50.

3. A geologic map of Arizona on the same scale as the base map, printed in many colors. It was issued in 1925, and is sold both mounted and unmounted for the same price as the topographic map.

The following unmounted Arizona map may be obtained from the U. S. Geological Survey, Washington, D. C., for $1.00.

A relief map of Arizona on the same scale as the base map, printed in various shades of brown, black, and blue. It was issued in 1925, and looks exactly like a photograph of a relief model of the State.

POSTAGE IS PREPAID ON ALL MAPS.

SERVICE OFFERED BY THE BUREAU

The Arizona Bureau of Mines will classify free of charge all rocks and minerals submitted to it, provided it can do so without making elaborate chemical tests. Assaying and analytical work is done at rates fixed by law, which may be secured on application.

The Bureau is always glad to answer to the best of its ability inquiries on mining, metallurgical, and geological subjects; and takes pride in the fact that its replies are always as complete and authoritative as it is possible to make them.

All communications should be addressed and remittances made payable to "The Arizona Bureau of Mines, University Station, Tucson, Arizona."

Vol. VI, No. 3   April 1, 1935

University of Arizona Bulletin

ARIZONA BUREAU OF MINES

G. M. Butler, *Director*

TREATING GOLD ORES

(Second Edition)

By

T. G. Chapman

ARIZONA BUREAU OF MINES, METALLURGICAL SERIES NO. 4

BULLETIN No. 138

PUBLISHED BY

University of Arizona

TUCSON, ARIZONA

University of Arizona Bulletin

Vol. VI, No. 3   April 1, 1935

HOMER LeROY SHANTZ, PH.D., Sc.D.................President of the University

STATEMENT OF MAILING PRIVILEGE

The University of Arizona Bulletin is issued semi-quarterly.

Entered as second-class matter June 18, 1921, at the post office at Tucson, Arizona, under the Act of August 24, 1912. Acceptance for mailing at special rate of postage provided for in Section 1103, Act of October 3, 1917, authorized June 29, 1921.

Vol. VI, No. 3

April 1, 1935

University of Arizona
Bulletin

ARIZONA BUREAU OF MINES

G. M. Butler, *Director*

TREATING GOLD ORES

(Second Edition)

By

T. G. Chapman

ARIZONA BUREAU OF MINES, METALLURGICAL SERIES NO. 4

BULLETIN No. 138

PUBLISHED BY
University of Arizona
TUCSON, ARIZONA

PREFACE

As is stated by the author of this bulletin, it was prepared not to enable the owners of relatively small gold mines to decide what process should be used in extracting the gold from their ores, but rather, to acquaint them with the various processes available and the advantages and disadvantages of each of them.

No treatment plant of any kind should ever be erected, no matter how small the cost, until the ore to be treated has been completely tested. Thousands of dollars may easily be saved by spending a few hundred dollars on such tests. No successful mining company ever erects a mill until exhaustive tests have been made, and they are just as essential when a small, "rawhide" plant is to be built.

A complete ore test is a very time-consuming job; it may easily take the entire time of one man for a month. When it has been completed, however, it is possible to decide just what process or combination of processes will yield the largest profit, what the percentage of extraction will be, the approximate cost of the plant, and the probable profit per ton of ore treated, assuming that the full tonnage that the plant is designed to handle is treated each day.

Not only should ore be thoroughtly tested before any considerable sum is spent on the erection of a treatment plant, but no such plant should be even considered until enough ore has been blocked out or otherwise developed to insure the operation of the plant for a long enough period to repay its cost and yield a satisfactory profit.

If no treatment plant of any considerable capacity were ever erected until sufficient ore to make the enterprise profitable was assured, and if the nature of all treatment plants were determined by careful, thorough ore tests, much of the risk connected with mining enterprises would be removed.

G. M. BUTLER.

April 1, 1935.

TABLE OF CONTENTS

PAGE

Preface 2

Introduction 5

Classification of methods of treating gold ores 6

Smelting of gold ores 6

Milling of gold ores 8

Amalgamation 8

      Advantages and disadvantages 11

      Amalgamating practice 11

Gravity stamps 11

Retorting 15

Melting and refining of retort bullion 16

Ball mills 17

Huntington mills 17

Arrastras 17

Amalgamating pans 20

Cost of recovering gold by amalgamating methods 20

Cyanidation 22

Gravity concentration 25

Flotation 26

ILLUSTRATIONS

Fig. 1.—Stamp battery 13

Fig. 2.—A retort 16

Fig. 3.—Arrastra 18

Fig. 4.—Amalgamating pan 19

TREATING GOLD ORES

By T. G. CHAPMAN

Metallurgist, Arizona Bureau of Mines; Professor of Metallurgy and Ore Dressing, University of Arizona.

INTRODUCTION

The Arizona Bureau of Mines receives many inquiries regarding methods for treating gold and silver ores. Due to the widespread interest in the treatment of these ores at this time, the similarity in the inquiries received, and the fact that many letters have been addressed to the Bureau by persons who desire replies in non-technical language, it was believed advisable for the Bureau to reprint Bulletin No. 133 entitled "Treating Gold Ores," published in April, 1932, for the use of persons not familiar with technical terms. It is obvious that a paper so written cannot cover the subjects included in more than a superficial manner, but it is believed that the Bureau can better serve the demand for such information by sending a bulletin rather than writing individual letters on the subject when requested to do so.

The writer wishes to acknowledge the kind assistance of Dean G. M. Butler, College of Mines and Engineering, University of Arizona, in reading the manuscript.

For those people who desire more detailed descriptions of gold milling processes, the following books are recommended: *Textbook of Ore Dressing*, by Richards and Locke, published by McGraw-Hill Book Co., New York. This book covers the subject of gravity stamps, amalgamation, gravity concentration, and flotation, and costs $5.50. *Manual of Cyanidation*, by Hamilton, published by the same company, describes the various methods used for cyaniding gold ores and costs $3.00.

It should also be noted that the United States Bureau of Mines has published Information Circulars which describe gold milling methods and costs as practiced in individual plants in the United States, Canada, and Mexico. These papers may be obtained free of charge by writing to the Director, U. S. Bureau of Mines, Washington, D. C. The numbers and titles of the circulars dealing with gold milling are as follows:

I. C. 6236—Milling Practice at the Alaska Juneau Concentrator.
I. C. 6408—Milling Methods and Costs at the Homestake Mine, Lead, S. Dak.
I. C. 6411—Milling Methods and Costs at the Spring Hill Concentrator, Helena, Mont.
I. C. 6433—Amalgamation Practice at the Porcupine United Gold Mines, Ltd., Timmins, Ont.
I. C. 6476—Milling Methods and Costs at the Argonaut Mill, Jackson, Calif.
I. C. 6541—Milling Methods and Costs of the Coniaurum Mines, Schumacher, Ont.
I. C. 6611—Small-Scale Placer Mining Methods.

CLASSIFICATION OF METHODS OF TREATING GOLD ORES

The common methods currently employed for the recovery of gold from ores derived from lode deposits may be classified as follows:
 1. Smelting
 2. Amalgamation
 3. Cyanidation
 4. Gravity concentration
 5. Flotation
One of these methods may be used alone or the methods may be combined, two or more being used in treating one ore. For example, after extracting part of the gold by amalgamation, a further recovery is sometimes profitably made by cyanidation, concentration, or flotation.

SMELTING OF GOLD ORES

Although examples of successful small-scale smelting operations applied to small tonnages of very high grade blanket, table, and flotation concentrates can be cited, the smelting method cannot usually be employed to advantage in treating average grade gold mine ores or concentrates in small-scale plants. In small-scale smelting operations, when applied to mine ores, the initial plant investment is heavy per ton of daily capacity and the operating cost is high. Since the initial plant investment is high, a considerable tonnage of high-grade ore should be developed before even considering small-scale smelting treatment. In order to obtain a high recovery of gold by smelting methods, it is necessary to use lead or copper ore in the furnace charge, and furthermore, the lead or copper ore must at times be prepared for the smelting operation by roasting and sintering preliminary treatments. For the reasons outlined, it has been customary for miners and prospectors operating on a small-scale basis to ship their ores to a custom smelting plant when the smelting method is to be used for the recovery of gold. The advantages to be gained by shipping rather than constructing a small smelting plant may be summarized as follows:

1. A large developed ore body is not a requirement.
2. Large initial plant investment is not necessary.
3. The purchase or mining of copper or lead ores is not necessary.
4. The risk involved in the proposition as a whole, when no smelter is erected, is not so great, due to the smaller investment necessary.

The advantages often cited in favor of small-scale smelting operations as compared to selling ores to custom smelting plants are first, a saving in freight charges, and, second, a saving in treatment charge. The first advantage is sound, and if any justification for small-scale smelting plants exists, it is to be found in an isolated locality where the charge for shipping mine ore is prohibitive, when the ore is refractory to milling processes, and when fuel, flux, lead or copper ores, and water are available at costs which leave a balance of profit on the operation as a whole. The second advantage, namely, that the treatment cost will be less, is not usually well founded. With present competition for ores by smelting plants and with the more efficient operation of large custom smelters as compared to small-size plants, it will usually figure to the seller's advantage to ship to those plants rather than consider the erection of a small plant.

In marketing gold ores to smelting plants, the seller should remember that the accurate sampling of a shipment is essential in order to avoid misunderstandings when settlement is made. The correct sampling of certain types of gold ores is not easy and at times is impossible without proper crushing equipment owing to the "spotty" character of such material. Again, in marketing gold ores, the shipper should send representative samples to all custom lead and copper smelting plants within reasonable shipping distances. The net returns from different plants will not always be approximately the same because one plant might need the waste material contained in the ore under consideration as a flux more than other plants. In such a case, a favorable smelting rate might be offered which would yield a higher net return to the seller. In summarizing the smelting treatment of gold ores, it can be stated that, for small properties within reasonable distances from custom lead and copper smelting plants, greater profit has often resulted in selling the material to the smelter rather than from the erection of expensive treatment plants, and, when using this method of disposing of gold ores, the shipper should undertake the proper sampling of the shipment, and, if proper crushing equipment is not available, should provide for a representative at the smelter during the sampling of his ore in order to prevent misunderstandings in settlements.

Since the publication of the first edition of this bulletin in April, 1932, the Mint price of gold has increased to $35 per troy ounce. Although the increase in the price of gold has permitted the treatment of lower grade ores and has increased the net returns from ores sold to custom smelting plants, it has also resulted in larger deductions, expressed in dollars, for the gold purchased by custom smelting plants. This condition has resulted in the existence of a possible margin between the total charge for smelting, including deductions, and the cost of treating high-grade concentrates at the milling plants. In the case of high-grade flotation, table, and blanket concentrates of gold mills, it is, under favorable conditions, feasible and more profitable to produce bullion for shipment to the United States Mint by small-scale smelting, amalgamating, or cyaniding treatments rather than to ship those products to custom smelting plants. The treatment of high-grade concentrate by the methods mentioned may not result in a high recovery of gold, but it should be noted that there is a rather wide margin on each ounce of gold so recovered even though it might be necessary to clean up and ship the residues remaining after such treatments.

MILLING OF GOLD ORES

The milling of gold ores is usually preferable to shipping to custom smelters providing the ore responds to milling treatment at a reasonable cost and also providing the ore body is of sufficient size to justify the expenditure necessary for constructing and equipping the milling plant. As already indicated, there are four methods available for milling gold ores, namely, amalgamation, cyanidation, gravity concentration, and flotation, and furthermore, as previously mentioned, these methods can often be combined to advantage.

AMALGAMATION

Amalgamation methods may be applied in various ways such as (1) gravity stamp mills which combine the crushing and amalgamating operation in a container called a stamp battery, (2) copper plates coated either with mercury or silver amalgam over which the ore pulp flows, (3) amalgamating pans, (4) arrastras which combine grinding of the ore with the amalgamating operation, and (5) sluices equipped with riffles, the purpose of using mercury in this connection being to catch fine gold. The last method is more applicable to the treatment of placer material than to the treatment of ore from lode deposits.

No matter what device is used in applying the amalgamation method, the basic principle is the same in all cases and it may be stated as follows: If clean, bright gold is brought in contact with clean, bright mercury, the two metals will alloy and form material called gold amalgam.

It should be noted that the surfaces of the gold and mercury must be clean and bright before the union of the two metals will take place. If the surface of the mercury is dark or tarnished, the gold, no matter how bright or clean, will not be caught by the mercury. Furthermore, if the mercury is broken up or subdivided into many small globules, the gold will not be caught. Mercury is rendered dark or coated or is broken up in the following ways: (1) it is tarnished by exposure to the air especially after being in use for a short period of time; (2) it is also tarnished and coated by uniting with base metals such as copper, lead, arsenic, or antimony (the source of these base metals is the ore treated or, in the case of copper, the amalgamating copper plate); (3) mercury is also coated by certain minerals at times found in gold ores, notably talc; (4) mercury is broken up into small globules by grease or oil derived from materials used in the mine or mill.

The methods ordinarily used to keep the surface of the mercury bright and clean are: (1) proper care in preventing grease or lubricating materials from coming in contact with the mercury and the use of lye, soda ash, or lime for cutting grease if present; (2) the use of a very small amount of sodium cyanide in removing the tarnish or stain; (3) the addition of a minute amount of metallic sodium to the mercury; (4) the use of common salt.

In using sodium cyanide for the removal of stains from copper amalgamating plates, it should be remembered that this substance is very poisonous even in minute quantities and, furthermore, cyanide dissolves gold. A piece of sodium cyanide about one inch in diameter if held in tongs and placed just above the stain for a few seconds with the ore pulp flowing over the plate, is sufficient to remove copper stains. The use of excessive amounts of cyanide should be avoided.

Although the addition of sodium to mercury has, in exceptional cases, proved of benefit in keeping the surface of mercury in good condition, its use is not recommended in general as it causes base metals to enter the amalgam which, after a short period of time, oxidize and cause the mercury to tarnish. In this way it gives more trouble than if it were omitted. If sodium is found to be beneficial it should be used in small amounts and with caution. Since sodium amalgam will be found difficult to purchase in good condition, due to the fact that it loses strength rapidly, two methods are given for its preparation, as follows:

Sodium amalgam is usually made by adding metallic sodium directly to mercury. The sodium should be cut into pieces about the size of the head of a common pin and these pieces added one by one to mercury until the sodium amalgam formed will just adhere to a clean nail. If the sodium does not readily react with the mercury, wet the end of a glass rod with water and touch the sodium resting on the surface of the mercury with the glass rod. The amalgam will then form. The amount of sodium necessary to add will, of course, depend upon the amount of mercury used and also on the condition of the mercury. Richards and Locke[1] state that one part of sodium to 2,000 parts of mercury were used

---

[1] Richards, R. H. and Locke, C. E., *Textbook of Ore Dressing*, McGraw-Hill Book Co., New York.

with good results in the treatment of Nova Scotia gold ores containing arsenic and talcose slate. The writer ordinarily uses from one to four parts of sodium to 2,000 parts of mercury and has found the "nail" test rather than a definite weighed amount of sodium to be the best rule for its use.

Since metallic sodium is not usually available in the field, another method which utilizes common salt and the electric current of an automobile battery for the making of sodium amalgam is described. In this latter method, mercury is placed in a glass container and covered with a solution of 3 to 5 ounces of common salt per quart of water. A wire from the positive pole of the battery is suspended in the salt solution and the negative wire is inserted into the mercury. A hollow glass rod placed with one end immersed in the mercury before adding the salt solution will serve to insulate the negative wire from the salt solution. The electrolysis of the salt solution results in the formation of sodium amalgam which is tested for strength by immersing a clean iron nail and noting when the amalgam is just active enough to adhere to the nail. Electrolysis for ten to fifteen minutes is usually sufficient for a few ounces of mercury, and correspondingly greater periods of time are required for large amounts.

The use of common salt for eliminating trouble caused by talcose materials has been suggested at times although no definite examples have been cited to the writer's knowledge. Salt, when used for such a purpose, is usually added intermittently, a shovelful being added to the ore pulp when found necessary.

Having considered the fact that mercury used for amalgamating gold must be clean and bright, and having described the various ways by which mercury becomes tarnished, coated, or subdivided into globules and the methods used in practice for keeping the mercury in good condition, the condition of the gold for successful recovery will next be described.

Coarse gold, in the free condition, is usually clean and bright especially after grinding of the ore, and, when in this condition, is readily recovered by amalgamation. Gold when present under the following conditions, however, cannot be amalgamated by ordinary methods:

1. Gold attached to or enclosed in quartz.
2. Gold contained in base metal sulphides, notably iron sulphide or pyrite.
3. So-called "rusty gold" which is believed to be free gold coated with iron oxide.

A fourth form of gold which is difficult to recover by amalgamation is finely divided gold. In mentioning coarse and fine gold, no definite size can be given, but any particle of gold readily visible to the naked eye, without straining the vision, might be roughly termed coarse gold. While fine gold is difficult to amalgamate, the operation may be assisted by impinging the gold on the surface of mercury. The impinging action is usually obtained by placing amalgamating plates within the stamp battery and thereby utilizing the splash, caused by the dropping of the stamps, to drop the ore pulp containing the gold onto the surface of the mercury or by breaking up the amalgamating plate that follows the stamp battery into a number of short sections, arranged as steps, the drop from step to step being about ½ inch. As the ore pulp drops a distance of ½ inch from step to step, the gold particles impinge upon the mercury coating of the plate, a condition which increases the amount of the finely divided gold particles recovered.

ADVANTAGES AND DISADVANTAGES OF AMALGAMATION

Due to the fact that in most ores a considerable portion of the gold is attached to quartz or associated with base metal sulphides, even after fine grinding, the loss of gold by the amalgamation method is considerable as compared to cyanidation, and for this reason, amalgamation is usually followed by either cyanidation, tables, or flotation in order to increase the recovery of gold in these forms. In general, cyanide treatment will recover part of the gold attached to quartz or associated with iron sulphide and tables or flotation will also partly recover the gold in these two forms. While it is a fact that amalgamation usually makes a low recovery as compared to other methods, it has the advantage that the product produced is gold bullion, readily and cheaply marketed. The amalgamation process also has the advantage of being easier to operate as compared to cyanidation and flotation and, while, as in other things, experience is necessary for efficient work, it takes less time, on the whole, for a man to become proficient in the practice of amalgamation than either cyanidation or flotation.

AMALGAMATING PRACTICE

Grinding of ores prior to amalgamating is usually done by gravity stamps, ball mills, Huntington mills, or arrastras.

GRAVITY STAMPS

Figure 1 is a sectional view of a stamp battery, from Richards' *Textbook of Ore Dressing*. Briefly, it consists of a battery box, bolted firmly to a heavy concrete foundation and fed from an ore bin by hand feeding or a mechanical ore feeder. The ore, previously crushed to between 3-inch and 1-inch sizes, is crushed further within the battery box by means of stamps raised by two-way cams and dropped by gravity through distances of 6 to 10 inches at the rate of ninety to one hundred drops per minute per stamp. A screen placed in front of the battery box confines the ore within the box until crushed to the desired size. Water, added at the rate of 4 to 7 tons per ton of ore treated, is fed with the ore. Mercury is usually added to the battery with the ore by means of a wooden spoon, the amount varying from 1 to 2 ounces per ounce of gold amalgamated. The addition of mercury to the stamp battery is of advantage to the small operator in that it catches part of the gold in the coarse condition within the battery. Plates within the battery utilize the impinging action caused by the splash of the ore pulp when the stamps drop which throws gold particles against the plates, a condition favorable, as previously mentioned, to amalgamating fine gold. The disadvantages of inside plates include the difficulty of keeping them in good condition, since they cannot be examined readily, and the scouring action of the ore pulp.

The rate of feeding ore to the stamps is controlled by the sound or feel of the blow struck by the stamps. A soft, mushy blow indicates over-feeding while a metallic blow with excessive rebound of stamps indicates under-feeding. Over-feeding of stamps should be especially avoided since very little crushing takes place within an over-fed battery and it takes considerable time to clear out a choked battery after feeding of ore has been stopped.

Occasionally a shoe becomes loosened and drops off the boss. The whole battery must then be shut down in order to refasten or replace the shoe. Soft pine wedges, which swell after wetting in service, are usually used for holding the shank of the shoe to the boss.

In order to avoid excessive whirling of the stamp, which is caused by friction with the cam, regular greasing of the cam is necessary, and, due to the harmful effects of grease upon mercury, this greasing must be done carefully and any excess of grease avoided. A rag rubbed on hard or medium-hard grease and then held against the contact surfaces of the cam, as it revolves in operation, will place on the cam a thin film of grease which is sufficient to prevent excessive whirling of the stamp.

The height of drop of the stamps is varied within certain limits by raising and lowering the tappet on the stamp stem. The number of drops of a given stamp per minute is limited by the height of drop and by the time consumed per drop. If more than one hundred drops per minute are attempted, there is danger of the tappet striking the arm of the cam and breaking it during the descent of the stamp.

When the ore pulp passes through the battery screen, it drops to a copper plate placed upon a slight slope. This copper plate is coated either with mercury or silver amalgam and a second catch of gold is effected on this plate.

Upon leaving the plate, the ore pulp passes through a trap for the recovery of mercury and amalgam which have been scoured off the plate by the pulp stream. The ore pulp is then, at times, as previously stated, treated by tables, cyanidation, or flotation in order to recover fine gold or gold attached to quartz or combined with base metal sulphides.

At certain time intervals, twelve to twenty-four hours, amalgam is removed from the plate, especially from the upper portion of the plate; it is first softened by adding more mercury and then scraped together with a rubber squeegee and removed. The plate

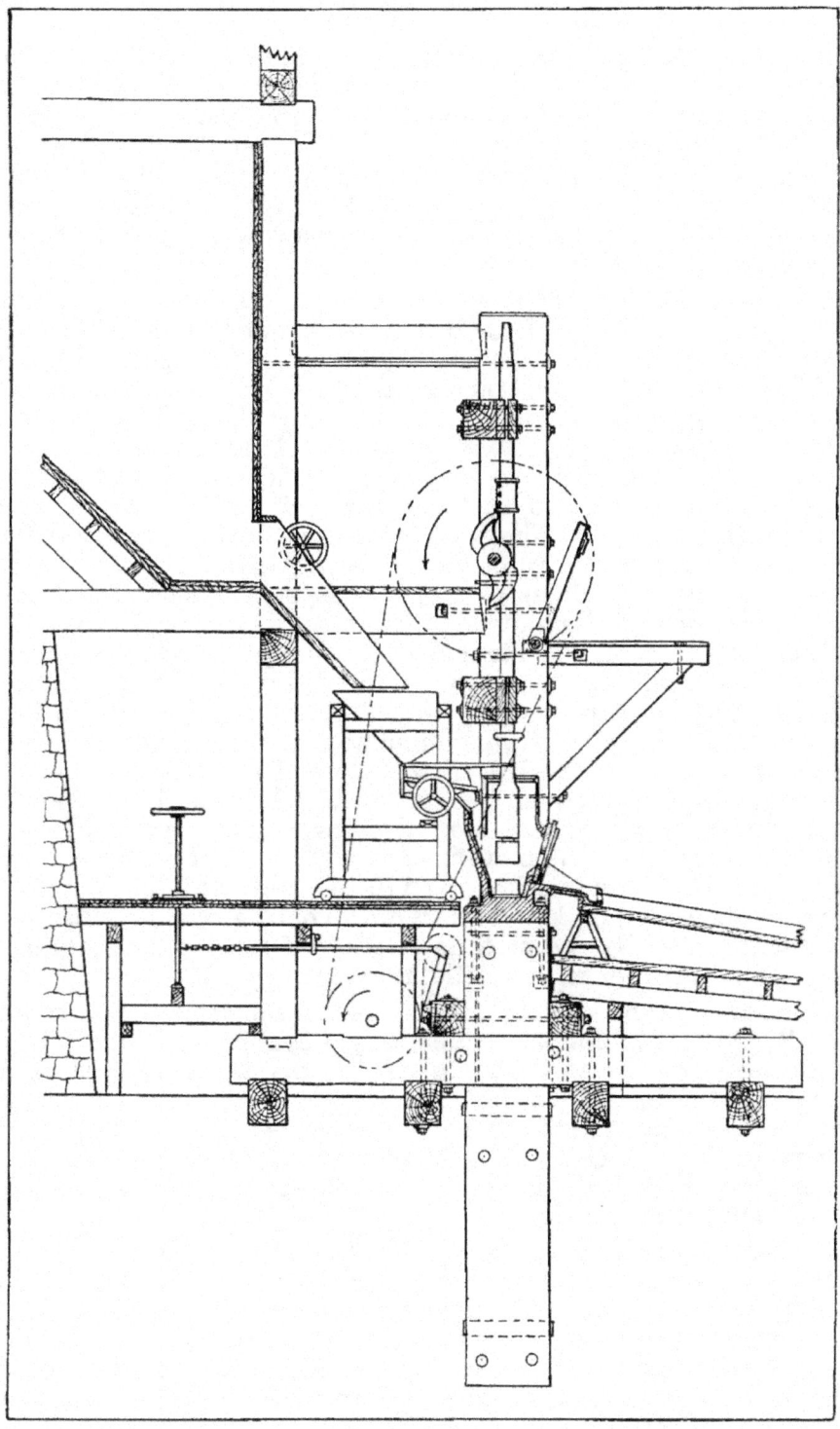

Figure 1.—Stamp battery. Taken from Richards' *Textbook of Ore Dressing*, McGraw-Hill Book Co., New York.

is then dressed by rubbing with a rag and any stains removed with a small amount of cyanide. During the operation of the stamps, stains are also removed from the plate when necessary, by means of cyanide as already indicated. A convenient method of adding mercury to plates is by means of a mercury "shaker" which is made by partially filling a 4-ounce, wide-mouthed bottle with mercury and tying a small piece of chamois or fine canvas over the mouth.

Although many materials have been used for amalgamating plates, soft, annealed, rolled-sheet copper, not less than 1/16 inch in thickness, is the most common plate material used. Richards and Locke[2] describe the Louis method of preparing new copper plates for service as follows: "Fine sand (sea sand if obtainable) is sprinkled on the plate, well moistened, and rubbed in with a block of wood until every portion of oxide is removed and the plate has a uniform red surface, care being taken at the same time not to scratch it. The sand is then washed off, and the plate dried and polished with fine emery paper folded over a block of wood. A perfectly clean, dry surface is thus produced. A mixture is then made of about ten parts sand to one part of coarsely pounded sal ammoniac; this mixture is dampened with water and clean, pure mercury is sprinkled into it by squeezing through canvas. The mixture is then rubbed over the plate with a piece of canvas or blanket when amalgamation will at once begin; more mercury must be sprinkled on the plate from time to time, and the rubbing continued until a uniformly bright, silvery surface is obtained. Each square foot of copper will require about ½ ounce of mercury. The plate is next washed well with water and kept until the following day. It will then probably be found that the plate is dulled and covered with a coating of a greenish-gray substance. Usually the plate is brightened with a dilute solution of cyanide, together with a little mercury."

The Louis method, as described, prepares new plates by removing all grease and copper oxide surface coatings and then puts a thin coating of copper amalgam on the plate. It has sometimes been found advisable to use a coating of silver amalgam rather than copper amalgam for new plates since silver amalgam will insure a better recovery of gold than copper amalgam when the latter is used on new plates. A convenient method of preparing silver amalgam follows.

Use sufficient silver to allow about ¼ ounce of silver per square foot of plate surface. Dissolve the silver in nitric acid (1.2 specific gravity). Remove the excess acid by evaporation, but do not bake the residue. Re-dissolve the silver nitrate in distilled water, using a small amount of nitric acid if found necessary to obtain complete solution of the silver. Dilute the solution to between 1,000 cc. and 2,000 cc. (1 to 2 quarts) and place strips of sheet

² Richards, R. H. and Locke, C. E., *Textbook of Ore Dressing*, p. 57, Mc-Graw-Hill Book Co., New York.

copper in the solution to precipitate the silver in a very finely divided form. After twelve to twenty-four hours remove the copper strips with attached silver to a separate glass container and dilute with one quart of distilled water. Heat almost to boiling, remove the copper strips and allow the fine silver to settle. Decant off the water and repeat the washing five to eight times to remove all copper. Finally, dry the silver and add mercury to it until the resulting amalgam can be easily molded in the hand but is not sufficiently liquid so that liquid mercury is oozing out of the mass (about three parts of mercury to one part of silver is required). After cleaning the plate by scouring as described by the Louis method, the silver amalgam is painted on the plate by rubbing the ball of amalgam back and forth sidewise (but not lengthwise) on the plate and finally smoothing the amalgam to a more or less uniform thickness with a clean brush or whisk broom.

At intervals varying from a week to a month, the stamp battery is shut down and, after removing the screen, shoes, and dies, the battery is cleaned out. At these clean-up periods, the amalgamating plate is scraped with putty knives and polished. After removing the accumulated amalgam and residue in the battery, repairs are made and the battery reassembled for another run.

The amalgam is usually recovered from the battery residue by panning, and, together with the amalgam recovered from the plate, is first cleaned by washing and then squeezed through chamois or other material. The mercury that passes through this filter is nearly free of gold and may be used again. The solid amalgam which is retained by the filter is ready for separation of the gold and mercury by retorting.

RETORTING

A retort for small-scale work is shown in Figure 2 and consists of an iron crucible equipped with a tight fitting cover held in place by a clamp. A hole is bored into the cover and a bent iron pipe for removing the mercury fumes is fitted into this hole. The discharge end of this bent pipe is surrounded by a water jacket through which cold water circulates and causes the mercury fumes to condense to liquid mercury. Upon heating the retort, the mercury contained in the gold amalgam vaporizes and enters the iron pipe where it is condensed to liquid and finally drops into a container partly filled with water, provided for the purpose. The gold remains in the retort.

The inside of the retort is usually coated with a very thin chalk or clay emulsion to prevent the gold bullion from sticking to the retort. A double thickness of newspaper also serves the same purpose. The cover, after being fastened in place by a clamp, is luted with chalk or clay emulsion. The retort should not be filled more than two-thirds to three-fourths full and the discharge pipe should barely dip beneath the surface of water so that all the fumes will condense and, when the retort cools, although water

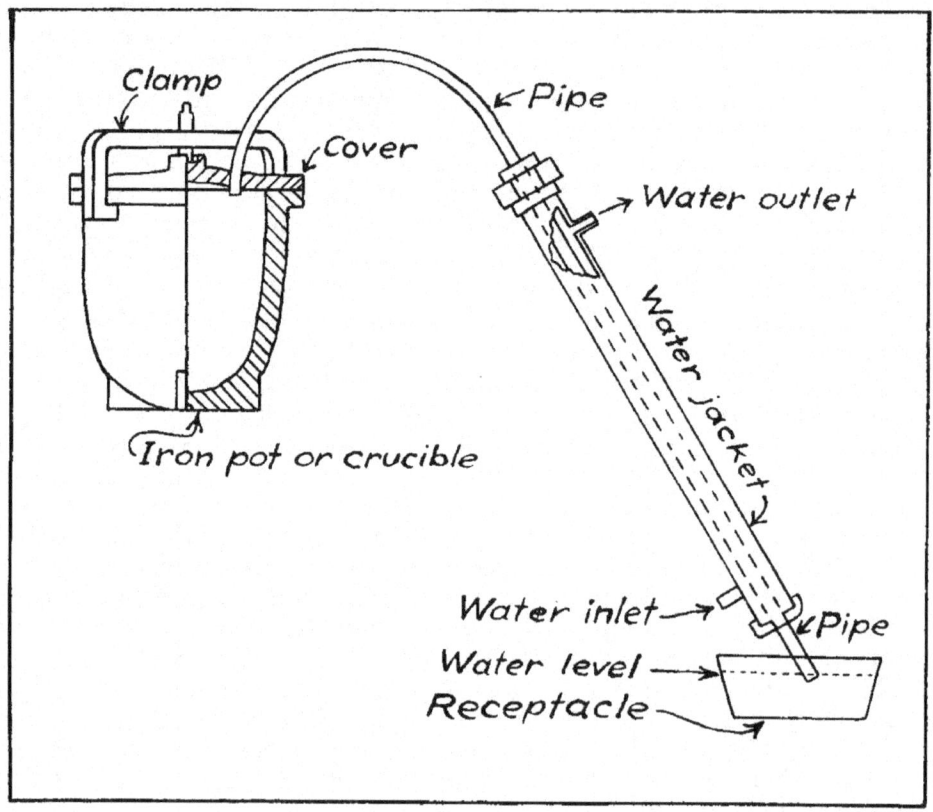

Figure 2.—A retort.

may be drawn into the pipe, water will not reach the crucible and cause it to explode.

It should be remembered that the fumes of mercury are poisonous, and for this reason, the retorting operation is best performed in the open, using a wood fire. The retort crucible should be allowed to cool before opening. If the operation is conducted indoors, special precautions must be taken to prevent the mercury fumes from concentrating in the atmosphere.

MELTING AND REFINING OF RETORT BULLION

The residue remaining in the retort after the removal of the mercury is usually placed in a graphite crucible and melted with fluxes. The fluxes commonly used for this purpose are niter, silica, sodium carbonate, and borax. After thorough melting and stirring, the contents of the crucible are usually poured into an iron mold previously coated with lubricating oil. The slag contains some gold and should be saved for re-treatment or sold to a smelter after a sufficient amount has accumulated.

BALL MILLS

Ball mills have replaced gravity stamps to a considerable degree during recent years in grinding ores prior to amalgamation in large-capacity plants especially when it is desired to follow amalgamation with very fine grinding for cyanidation. In small-capacity plants, however, the gravity stamp mill retains favor, especially when treating hard ore, because it is easier to regulate small-capacity gravity stamp mill units to the desired capacity than is possible when ball mills are used. Gravity stamps are also convenient amalgamators whereas certain difficulties, including the flouring of mercury when placed in the ball mill and the segregation of coarse gold in the ball mill, make the operation of amalgamation somewhat more difficult in ball mills than in stamps. Ball mills, however, can be used to advantage if proper precautions are taken for amalgamation and the choice between stamps and ball mills will depend largely upon existing conditions.

HUNTINGTON MILLS

Huntington mills have been used in the past for crushing ores prior to amalgamation. A mill of this type consists of rollers with vertical axles, which operate against a circular die or ring in the rim of a pan. Crushing is facilitated by centrifugal force which drives the rollers against the ring die. The shipping weight of this mill per ton of daily capacity is less than for gravity stamps and this fact is the chief point in favor of such a mill. The rollers, however, do not usually wear smoothly, and the rollers then pound against the die. Although the machine is a good amalgamator, it is not in favor at present owing to the high operating and maintenance charges as compared to stamps and ball mills.

ARRASTRAS

The arrastra is a device for doing the work of a stamp mill, but most of the materials used for its construction are available in the vicinity of the mine. A drawing of an arrastra is presented in Figure 3. Richards and Locke, in their *Textbook of Ore Dressing,* describe arrastras in general as follows: "This mill consists of a circular pavement from 6 to 20 feet in diameter, with a retaining wall around it and a step in the center. Upon the step stands a vertical, revolving spindle or shaft, and from the spindle extend horizontal arms to which large boulders, called drag-stones, are attached by chains. The boulders are dragged around the circle by arms and crush the ore by a true grinding action.

"The arms number from two to eight, usually four. The drag-stones vary from two to twelve, commonly four; they weigh from 80 to 2,000 pounds, averaging about 300 pounds. Holes are drilled in the stones, plugged with dry wood, and the eye rings are driven into these plugs. They are placed so that the stone shall slide on its largest surface and forward of the center of gravity so that

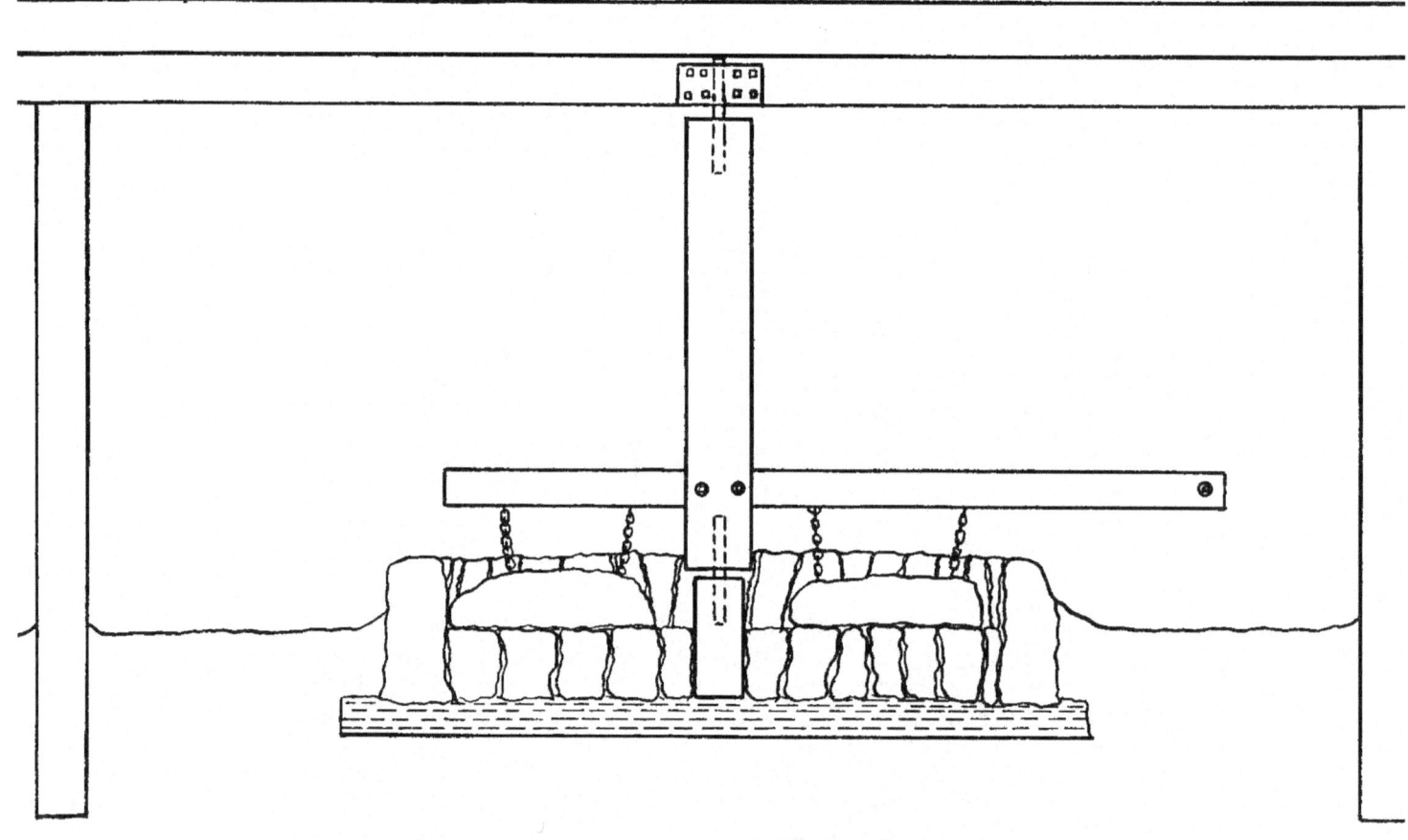

Figure 3.—Arrastra.

the front edge of the stone may be lifted sufficiently to slide over the coarsest of the ore during the early stage of grinding.

"To prevent leakage of quicksilver, the pavement is built upon a clay or concrete foundation which is always wider than the pavement. The latter is almost 1 foot thick, of granite, basalt, or flinty quartz, a rough-grained rock being preferred. The joints are filled with fine tailings, or, better, with cement. The retaining wall, 2 to 4 feet high, is made of stones laid in cement, of wooden staves bound with iron hoops, or is merely a clay bank. It has a gate or a series of plug holes for discharging the pulp and, sometimes, screen discharges for continuous work.

"The speed is four to eighteen revolutions per minute, usually ten to fourteen for power arrastras. Small arrastras are driven by a horse or mule attached to an extension of one of the arms, the animal walking around the circle. Large arrastras are driven by a horizontal water wheel, suspended from cross arms separate from the dragging arms and extending outside the retaining wall, or they are driven by a shaft with beveled gears. One long shaft may in this way connect several arrastras with a single driving engine.

"It is used as a fine grinder and amalgamator with both gold and silver ores, and is fed with material seldom above ¾ inch in diameter, often much below. It is used where cheapness, both of installation and of running, is essential and, at the same time, small capacity is not objectionable, for example, in regions remote from supplies. It is often used for re-treating tailings of gold mills, chiefly by lessees."

Arrastras are usually operated intermittently. A batch of ore is added with sufficient water to make a thick pulp when the ore has been ground. The mercury is usually added at the last stage of the grinding operation. After sufficient time has elapsed for amalgamation of the gold, the contents of the arrastra, including the amalgam, are removed and the amalgam is recovered by panning or in rockers. If sufficient water is available, the charge may be removed from the arrastra by sluicing and passed through traps for the recovery of the amalgam.

AMALGAMATING PANS

The amalgamating pan shown in Figure 4 is a device for cleaning up the residue of a stamp battery or for cleaning up black sands from placer work. A rotating barrel is often used for the same purpose.

The grinding pan comprises a cylindrical container into which the material to be amalgamated is placed with water and mercury. The pulp is mixed and stirred by means of revolving framework called a muller. The pulp should be sufficiently thick to maintain finely divided mercury globules in suspension but not thick enough to prevent movement of the globules in the pulp. After a few hours of stirring, the mixture is removed and the amalgam separated by pans or rockers.

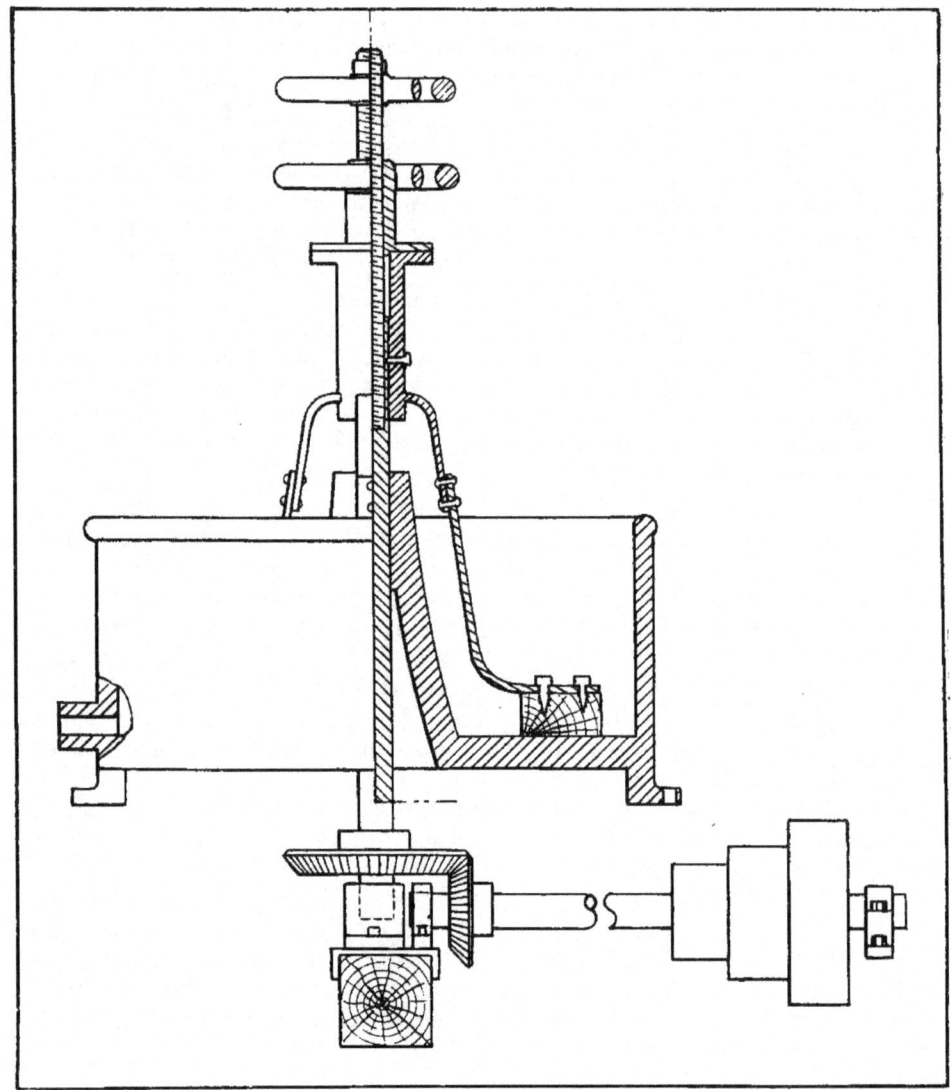

Figure 4.—Amalgamating pan.

COST OF RECOVERING GOLD BY AMALGAMATION
METHODS

It is difficult to give costs of amalgamating gold ores that
would be useful to small operators since operating milling costs
in small plants depend to a large extent upon the character of the
ore treated and operating conditions. Operating conditions vary
greatly in such factors as per cent of operating time obtained by

the mill, cost of water, power, and labor, and accessibility of the mill to supplies.

For the purpose of indicating cost of amalgamating gold ores on a small-capacity basis, the following costs are given for a 25-ton capacity amalgamation mill in Canada. The figures have been taken from United States Bureau of Mines I. C. 6433 (Amalgamation Practice at Porcupine United Gold Mines, Ltd., Timmins, Ont.) by R. A. Vary, March, 1931.

OPERATING COST PER TON OF ORE TREATED

	Labor	Power	Supplies	Total
Crushing	$0.214	$0.135	$0.035	$0.384
Grinding	0.429	0.180	0.162	0.771
Classifying, screening conveying and refining....	0.429	0.090	0.107	0.626
Miscellaneous	....	....	0.070	0.070
Totals	$1.072	$0.405	$0.374	$1.851

As previously mentioned, these figures are to be taken only as an indication of cost per ton in a well-operated 25-ton capacity plant. For a mill less efficiently operated, less accessible to supplies, or treating a more refractory ore, the cost would be greater. It should also be noted that the costs given are operating costs only and do not take into consideration writing off the first cost of building and equipping the mill. For small ore bodies, this latter cost is considerable and is at times the controlling factor between profit and loss on the undertaking as a whole.

Richards and Locke give combined crushing, stamping, amalgamating, and concentrating operating costs from data collected by the General Engineering Company of Salt Lake City for a period comparable to present unit power, labor, and supply costs. This tabulation is given to show the effect of plant capacity upon operating costs and indicates the disadvantage to which the small operator is put in this respect.

Capacity, tons per 24 hours	Location	Description of treatment	Cost per ton
30	Calif.	Crushers, gravity stamps, and amalgamation	$1.92
50	Calif.	Crushers, gravity stamps, amalgamation, gravity concentration	1.07
300	Calif.	Crushers, gravity stamps, amalgamation	0.22
1000	Alaska	Crushers, gravity stamps, gravity concentration, and amalgamation	0.24

CYANIDATION

The cyanide method of recovering gold from ores is based upon the principle that dilute solutions of sodium or potassium cyanide (sodium cyanide is commonly used) will dissolve gold from crushed or ground ores under certain conditions. After the gold has been dissolved, the solution is separated from the ore residue and the gold recovered from it by means of metallic zinc shavings or dust.

The cyanide method, in contrast to amalgamation, flotation, and gravity concentration methods, is distinctly a chemical method and cannot be used to advantage without chemical control of cyanide quantities, alkalinity, and assays of solutions and residues.

There are two general methods of applying the cyanide process to gold ores, namely: the percolation method and the agitation method. The size to which the ore must be crushed or ground prior to treatment in order to obtain satisfactory extraction of the gold within reasonable time periods largely governs the amount of slimes produced and is usually, therefore, the determining factor of deciding which of these two methods is to be used in small-scale operations.

In the percolation method, as applied to the entire ore, the material is crushed to between ¼ inch and about 10-mesh and placed in vats, usually made of redwood and equipped with false bottoms of cocoa matting and canvas, through which the solution, but not the ore, may penetrate. The cyanide solution, ranging in strength from 0.25 to 0.05 per cent, is added to the tank and slowly allowed to percolate through the ore charge. After sufficient time has been allowed for satisfactory solution of the gold, the pregnant gold solution is clarified and, in small-plant operations, is usually passed through boxes filled with zinc shavings to precipitate the gold. After the tank containing the ore residue has drained, barren solutions (solutions after passing through the zinc boxes) are used as washes and, finally, fresh water is added to remove as much as possible of the dissolved gold remaining in the residue. Lime is usually added to the crushed ore before applying the cyanide solution in order to neutralize any acidity in the ore and thereby save the expensive cyanide salt which is consumed by acid in the absence of lime.

In the agitation method, the ore is finely ground, usually in a cyanide solution, and, after grinding, the thick ore pulp is agitated in tanks by means of mechanical stirrers or compressed air for a period of time sufficient to obtain satisfactory solution of the gold. The gold-bearing solution is then separated from the solid ore residue by repeatedly diluting and decanting the liquid or by filtering. The gold is then precipitated from the clarified solution by metallic zinc.

The first method described, the percolation method, has two definite advantages over the second or agitation method, especially for the small operator, namely, lower first cost of plant and lower operating cost. The lower first cost results from the coarser crushing used in this method and the absence of mechanical agitators, filtering equipment, and the necessary power plant equipment to operate them. The lower operating cost is due to coarser crushing, elimination of power for agitation and filtration, and the simplicity of the process as compared to the agitation method.

The disadvantages of the percolation method as compared to the agitation method are lower extraction of the gold and longer periods of time usually required to extract the gold.

The percolation method should be thoroughly tested before installing, not only for such information as per cent recovery of gold, consumption of cyanide, time required for satisfactory gold recovery at size of crushing decided upon, but also for satisfactory percolation rate of solutions through the ore charge when using the same depth of ore charge that is planned for actual operation. This latter precaution is very important since many small-size plants have failed to operate successfully due to the fact that these preliminary tests were not made. It should be noted that, if a solution percolates satisfactorily through a column of crushed ore 6 to 12 inches high in a laboratory percolator, it does not necessarily follow that solutions will percolate through ore columns 3 to 5 feet high at a satisfactory rate in actual operations. Resistance to the percolation of solutions depends upon the depth of ore charges, the size of crushing, and the amount of slimes contained in the charge. Before deciding upon the installation of percolation equipment, a test should be made in which the depth of ore charge, the size to which the ore is crushed, and the amount of slimes are the same, as nearly as possible, as they will be in actual operations.

The agitation method, as indicated, possesses the advantages of higher recovery and shorter time of treatment as compared to the percolation method, owing to finer grinding used. Its high initial and operating costs are, however, serious obstacles to its installation for small-plant operations. In large-scale cyanide operations where maximum recovery is the governing factor, the ore must be finely ground, and the agitation method is then usually required since cyanide solutions will not percolate, as a rule, through finely ground ore beds. In small-scale operations, the first cost of the plant is at times the governing factor, and, therefore, in small-scale operations, it is sometimes more profitable to treat the coarsely crushed ore by percolation and accept the lower recovery than to attempt to extract more of the gold by a plant too expensive for the size of the ore body.

It is not possible to determine, without testing, whether an ore is or is not amenable to cyanide treatment. In general, ores containing finely divided metallic gold are more amenable to straight cyanidation than ores containing coarse gold since the time required to dissolve a given quantity of fine gold is less than that required to dissolve the same amount of coarse gold. Amalgamation can, however, be used prior to cyanidation and the coarse,.

free gold recovered at low cost by amalgamation, leaving the fine, included, and attached gold to be partly saved by cyanidation.

Besides the size of gold particles in ores, other characteristics of ores influence the amenability of ores to cyanide treatment. Among the most important of these characteristics, the following are briefly noted.

Gold, attached to or included in quartz or iron pyrite, is dissolved by cyanide providing the solution reaches the gold and sufficient time is available. Gold lost by the amalgamation method, due to its presence in these forms, is, therefore, partly recovered by cyanidation, and it is for this reason, chiefly, that recoveries of gold by cyanidation are usually higher than by amalgamation.

Sodium cyanide is a relatively expensive chemical reagent costing, at present, from sixteen to twenty cents per pound. The consumption of this reagent is, therefore, not only important as affecting cost of treatment, but is at times prohibitive. Cyanide is lost mainly in two ways: first, chemically, by reacting with various impurities contained in the ore, and, second, mechanically, due to imperfect washing of ore residues. Acids derived from organic matter and from oxidation of pyrite consume cyanide, but this loss is largely prevented by adding a cheaper alkali, lime or soda ash, to the crushed or ground ore prior to the addition of cyanide. The lime or soda ash reacts with the acid and thereby saves the cyanide. Copper minerals, especially oxidized copper minerals and iron compounds (ferrous) are the most common cyanide consuming impurities that lead to high and, at times, prohibitive cyanide consumptions. Although examples can be given of large plants operating successfully with these impurities present in the ore, it requires considerable skill and high plant investment to use the cyanide method profitably when appreciable amounts of these impurities are present in the ore.

An important condition that is encountered in cyanide operation is the fouling of solutions. This condition is caused by the accumulation of impurities in the solutions as they are used over and over again. The solutions may also deteriorate if certain impurities in them consume oxygen since gold will not be dissolved unless oxygen is present in the solutions. The chief effect of the fouling of solutions is either to reduce the amount of gold dissolved from the ore or to increase the time necessary for dissolving the gold. Both of these effects are serious in plant operation, and advance information as to the probable results of fouling can only be obtained by small-scale, continuous laboratory tests whereby the laboratory solutions are allowed to foul.

The gold precipitate produced by contact of the gold-bearing solutions with metallic zinc, although impure, can be marketed directly to a smelting plant. It is usually better, however, to remove most of the zinc that is mixed with the precipitate and then to dry and melt the material with proper fluxes in order to produce gold bullion bars which can be sold directly to the United

States Mint. A large amount of the zinc that is mixed with the gold precipitate may be removed by washing the precipitate on screens if zinc shavings were used for the precipitation operation. The screen treatment may be followed by sulphuric acid to remove additional zinc or the sulphuric acid treatment may be used directly if zinc dust has been used to precipitate the gold. The washed, cleaned precipitate is then nearly dried and melted with sodium carbonate, silica, and borax fluxes.

As already indicated, the cost of cyanidation depends upon the method selected and upon the character of the ore treated. Straight cyanidation by the percolation method is a cheap process when applied to a clean ore, but few ores can be treated by percolation alone since fine crushing of the ore is usually necessary for satisfactory recovery and this fine crushing in turn produces a product which contains too much fine material to be handled by this method. The percolation process has often been used on the coarser sand portion of an ore after comparatively fine grinding while the fine slimes are treated by the agitation cyanide method or by flotation. The Homestake Mining Company, in 1929, crushed, ground, amalgamated, and cyanided the sand by percolation and the slimes in filters for a total operating cost of $0.503[3] per ton of ore. The total cost was distributed as follows: $0.038 for coarse crushing, $0.276 for grinding and amalgamation and $0.189 for cyanidation. The efficient Homestake operation, conducted on a basis of about 4,000 tons per day, cannot of course be duplicated in a small plant as far as cost is concerned. Under favorable small-plant operating conditions, the operating cost for the percolation method may be as low as $0.50 to $0.75 per ton, and, under less favorable conditions, it might run up to from $2.00 to $3.00 per ton or higher. The operating cost for the slime agitation process, when operated on a small-capacity basis, will range from $3.00 to $6.00 per ton of ore treated.

GRAVITY CONCENTRATION

Although the gravity concentration method is often used without the aid of accessory methods in recovering gold from placer deposits, it is not often so used in the treatment of ores derived from lode deposits. In the treatment of the latter material, gravity concentration is more often used as an accessory method to recover part of the gold which escapes amalgamation, cyanidation, or flotation. Wet concentrating tables and vanners are commonly used for this purpose and many types are available. Some of them have stationary, sloping surfaces covered with canvas, blankets, cement, or riffles. Others are provided with movable decks which are either given a rapid forward and backward motion or, in the case of vanners, a more gentle sidewise or lengthwise, swaying motion.

---

[3] Clark, A. J., "Milling methods and costs at the Homestake Mines, Lead, S. Dak.," I. C. 6408, U. S. Bureau of Mines, Feb., 1931.

Tables take advantage of the facts that gold particles are not carried along by a current of water so rapidly as are the other constituents of the ore, and they settle more rapidly than other constituents of the ore in a liquid ore pulp and, therefore, find their way to the bottom of the pulp bed. While settling will take place without motion of the table deck, it is usually desirable to move the deck since the separation is then hastened and the treatment is continuous. In the case of very fine gold particles, however, motion of the table interferes with the recovery of gold, and for this reason, stationary canvas- or blanket-covered tables will effect a higher recovery of fine gold than movable tables of the Wilfley or Deister types. The capacities of stationary tables are, however, much less than movable tables and they operate intermittently, whereas the movable decks operate continuously, as stated. Intermittent operation means that the table is fed for a certain period during which time the pores of the canvas or blanket become filled with the minerals desired to be recovered; the feed is then stopped and the values removed after which the operation is repeated.

The moving power of a current of water is a very important feature of any wet concentrating table. The water which flows down the gentle slope of the table or vanner deck removes the waste material while the heavier minerals are retained by riffles, canvas, or blankets.

The usual cost of small-scale table and vanner treatment, not including the cost of crushing, grinding, screening, or other accessory operations, ranges from $0.05 to $0.25 per ton and depends largely on capacity handled and cost of water.

FLOTATION

The use of flotation methods in the treatment of gold ores is a comparatively recent development and offers considerable promise to small operators in that the first cost of the plant and the operating cost should be well under the figures required for a fine grinding agitation type of cyanide plant. Although the recovery of gold made by a flotation plant will not usually be so high as in the competing fine grinding cyanide plant and although the flotation concentrate is at times shipped to a smelting plant whereas gold bullion with its cheap marketable feature is obtained in a cyanide plant, nevertheless the lower first and operating costs of a flotation plant will, it is believed, in many instances, more than offset the advantages of cyanidation. Again, it is at times possible to cyanide the flotation concentrate produced and, when this is the case, one of the advantages of direct cyanidation is eliminated.

In the flotation process, the ground ore, mixed with water, is brought in contact with bubbles of air. The gold particles, possessing a bright, metallic surface, tend to adhere to the air bubbles whereas the earthy, waste constituents of the ore, having a dull, non-metallic luster, have less tendency to cling to air bubbles. The reagents usually used for the floating of gold, consist of lime, a collector, and a frother. The amount of lime needed varies for different ores and can be determined only by testing. The collector reagent increases the adhesion of the gold particles to the air bubbles. The results of experimental work done by the writer show that amyl xanthate and sodium aerofloat are usually satisfactory collectors in the treatment of gold ores. These reagents are used in small amounts, from 0.03 to 0.20 pound per ton of ore treated, and recent work by others indicates that smaller quantities at times produce better results. The frother is added to prevent the bubbles from breaking and, for gold work, steam-distilled pine oil, added at the rate of 0.05 to 0.10 pound per ton, has been found satisfactory.

There are many types of machines on the market for the treatment of gold ores and the chief difference in the principles of these machines is the manner of supplying the air bubbles. The type usually used for the treatment of gold ores is the mechanical agitation machine with or without sub-aeration. Air is supplied to the straight mechanical agitation type machine by beating it into the liquid pulp by the use of rapidly revolving arms or paddles. In sub-aeration mechanical machines, additional air is usually provided by a blower or other means, at a low pressure. The ore pulp, with reagents, is thoroughly mixed with the air in the agitation compartment of the machine and then discharged into a compartment for settling. The bubbles of air, with gold particles attached, form a froth which rises to the surface of the ore pulp from which position it is removed.

The second important type of machine is the pneumatic which receives all the air used from a low pressure blower. Some pneumatic machines are provided with a canvas mat at the bottom through which the air from the blower must pass before mixing with the ore pulp. The pneumatic machines that are not provided with mats are called either "matless" or air-lift type machines.

Results of comparative tests of the mechanical and pneumatic type machines, when operating with gold ores, have not been published, to the writer's knowledge. There is, however, a feeling among mill men that the mechanical type of cell is preferable for the treatment of gold ores, and tests made by the writer show that this type of cell is satisfactory for gold ores. As a rule the power cost for operating the mechanical type cell is higher than that for the pneumatic cell but, due to more efficient mixing of reagents, the mechanical cell usually is more economical in cost of reagents and in some instances, possibly, is superior from the standpoint of recovery.

With small plants operating on gold ores, it is believed that tables operated as scavengers, following flotation, will be found valuable. The operation of a table, as described, does not greatly increase the operating cost and it will serve as a pilot machine for observation of flotation results and will, in many instances, recover an appreciable percentage of the total gold.

In the flotation treatment of ores containing some coarse gold, amalgamation may be useful. From the results of tests made by the writer, amalgamation, if carried out, should be used before adding the flotation reagents to the ore as these reagents have been found to have a harmful effect upon the mercury.

In large flotation plants, the operating cost, including crushing, grinding, and all accessory operations, varies from $0.18 to $1.25 per ton of ore treated. With small plants this cost will vary from $1.00 to $2.50 per ton of ore treated, and the maximum figure given may be exceeded if the per cent of operating time is not maintained at a satisfactory figure.

In order to show the distribution of operating costs and metallurgical results of a 150-ton capacity mill, operating on gold ore by flotation methods, the following figures, recently published by the U. S. Bureau of Mines, are given for the Spring Hill Concentrator located near Helena, Montana.

SUMMARY OF MILLING COSTS AT THE SPRING HILL CONCENTRATOR, JULY 1, 1929 TO MARCH 3, 1930.
Cost per ton of ore milled*

	Labor	Super-vision	Supplies	Reagents	Power	Total
Crushing and conveying	$0.062	$0.001	$0.050		$0.049	$0.162
Grinding	.100	.001	.072		.117	.290
Flotation	.108	.001	.055	$0.063	.031	.258
Disposal of tailings	.048	.001	.003		....	.052
Water	....	.001	.001		.010	.012
Repairs and maintenance	.102	.001	.130		....	.233
Total	$0.420	$0.006	$0.311	$0.063	$0.207†	$1.007

*Grant, L. A., "Mining Methods and Costs at the Spring Hill Concentrator of the Montana Mines Corporation, Helena, Mont.," I. C. 6411, U. S. Bureau of Mines, March, 1931.
†Power costs $0.008 per kilowatt hour.

METALLURGICAL DATA AT SPRING HILL CONCENTRATOR, JULY 1, 1929 TO MARCH 31, 1930.

Heads assay..$6.46
Tailings assay..$1.22
Total ore concentrated, tons..40,930
Days operated..270
Operating time per day, hours..24
Operating time, per cent..98
Average ore concentrated per 24 hours, tons...................153
Recovery of gold, per cent...82
Concentration ratio...12 to 1
New water consumption per ton of ore concentrated, gallons...................104
Ball consumption per ton of ore concentrated, pounds.............1.30
Reagent consumption, pounds per ton of ore milled
    Pine oil..0.075
    Potassium ethyl xanthate.......................................0.109
    Aerofloat 0.103

GOLD RUSH BOOKS

OREGON, USA

www.GoldMiningBooks.com

Books On Mining

Visit: www.goldminingbooks.com to order your copies or ask your favorite book seller to offer them.

Mining Books by Kerby Jackson

Gold Dust: Stories From Oregon's Mining Years - Oregon mining historian and prospector, Kerby Jackson, brings you a treasure trove of seventeen stories on Southern Oregon's rich history of gold prospecting, the prospectors and their discoveries, and the breathtaking areas they settled in and made homes. 5" X 8", 98 ppgs. Retail Price: $11.99

The Golden Trail: More Stories From Oregon's Mining Years - In his follow-up to "Gold Dust: Stories of Oregon's Mining Years", this time around, Jackson brings us twelve tales from Oregon's Gold Rush, including the story about the first gold strike on Canyon Creek in Grant County, about the old timers who found gold by the pail full at the Victor Mine near Galice, how Iradel Bray discovered a rich ledge of gold on the Coquille River during the height of the Rogue River War, a tale of two elderly miners on the hunt for a lost mine in the Cascade Mountains, details about the discovery of the famous Armstrong Nugget and others. 5" X 8", 70 ppgs. Retail Price: $10.99

Oregon Mining Books

Geology and Mineral Resources of Josephine County, Oregon - Unavailable since the 1970's, this important publication was originally compiled by the Oregon Department of Geology and Mineral Industries and includes important details on the economic geology and mineral resources of this important mining area in South Western Oregon. Included are notes on the history, geology and development of important mines, as well as insights into the mining of gold, copper, nickel, limestone, chromium and other minerals found in large quantities in Josephine County, Oregon. 8.5" X 11", 54 ppgs. Retail Price: $9.99

Mines and Prospects of the Mount Reuben Mining District - Unavailable since 1947, this important publication was originally compiled by geologist Elton Youngberg of the Oregon Department of Geology and Mineral Industries and includes detailed descriptions, histories and the geology of the Mount Reuben Mining District in Josephine County, Oregon. Included are notes on the history, geology, development and assay statistics, as well as underground maps of all the major mines and prospects in the vicinity of this much neglected mining district. 8.5" X 11", 48 ppgs. Retail Price: $9.99

The Granite Mining District - Notes on the history, geology and development of important mines in the well known Granite Mining District which is located in Grant County, Oregon. Some of the mines discussed include the Ajax, Blue Ribbon, Buffalo, Continental, Cougar-Independence, Magnolia, New York, Standard and the Tillicum. Also included are many rare maps pertaining to the mines in the area. 8.5" X 11", 48 ppgs. Retail Price: $9.99

Ore Deposits of the Takilma and Waldo Mining Districts of Josephine County, Oregon - The Waldo and Takilma mining districts are most notable for the fact that the earliest large scale mining of placer gold and copper in Oregon took place in these two areas. Included are details about some of the earliest large gold mines in the state such as the Llano de Oro, High Gravel, Cameron, Platerica, Deep Gravel and others, as well as copper mines such as the famous Queen of Bronze mine, the Waldo, Lily and Cowboy mines. This volume also includes six maps and 20 original illustrations. 8.5" X 11", 74 ppgs. Retail Price: $9.99

Metal Mines of Douglas, Coos and Curry Counties, Oregon - Oregon mining historian Kerby Jackson introduces us to a classic work on Oregon's mining history in this important re-issue of Bulletin 14C Volume 1, otherwise known as the Douglas, Coos & Curry Counties, Oregon Metal Mines Handbook. Unavailable since 1940, this important publication was originally compiled by the Oregon Department of Geology and Mineral Industries includes detailed descriptions, histories and the geology of over 250 metallic mineral mines and prospects in this rugged area of South West Oregon. 8.5" X 11", 158 ppgs. Retail Price: $19.99

Metal Mines of Jackson County, Oregon - Unavailable since 1943, this important publication was originally compiled by the Oregon Department of Geology and Mineral Industries includes detailed descriptions, histories and the geology of over 450 metallic mineral mines and prospects in Jackson County, Oregon. Included are such famous gold mining areas as Gold Hill, Jacksonville, Sterling and the Upper Applegate. **8.5" X 11", 220 ppgs. Retail Price: $24.99**

Metal Mines of Josephine County, Oregon - Oregon mining historian Kerby Jackson introduces us to a classic work on Oregon's mining history in this important re-issue of Bulletin 14C, otherwise known as the Josephine County, Oregon Metal Mines Handbook. Unavailable since 1952, this important publication was originally compiled by the Oregon Department of Geology and Mineral Industries includes detailed descriptions, histories and the geology of over 500 metallic mineral mines and prospects in Josephine County, Oregon. **8.5" X 11", 250 ppgs. Retail Price: $24.99**

Metal Mines of North East Oregon - Oregon mining historian Kerby Jackson introduces us to a classic work on Oregon's mining history in this important re-issue of Bulletin 14A and 14B, otherwise known as the North East Oregon Metal Mines Handbook. Unavailable since 1941, this important publication was originally compiled by the Oregon Department of Geology and Mineral Industries and includes detailed descriptions, histories and the geology of over 750 metallic mineral mines and prospects in North Eastern Oregon. **8.5" X 11", 310 ppgs. Retail Price: $29.99**

Metal Mines of North West Oregon - Oregon mining historian Kerby Jackson introduces us to a classic work on Oregon's mining history in this important re-issue of Bulletin 14D, otherwise known as the North West Oregon Metal Mines Handbook. Unavailable since 1951, this important publication was originally compiled by the Oregon Department of Geology and Mineral Industries and includes detailed descriptions, histories and the geology of over 250 metallic mineral mines and prospects in North Western Oregon. **8.5" X 11", 182 ppgs. Retail Price: $19.99**

Mines and Prospects of Oregon - Mining historian Kerby Jackson introduces us to a classic mining work by the Oregon Bureau of Mines in this important re-issue of The Handbook of Mines and Prospects of Oregon. Unavailable since 1916, this publication includes important insights into hundreds of gold, silver, copper, coal, limestone and other mines that operated in the State of Oregon around the turn of the 19th Century. Included are not only geological details on early mines throughout Oregon, but also insights into their history, production, locations and in some cases, also included are rare maps of their underground workings. **8.5" X 11", 314 ppgs. Retail Price: $24.99**

Lode Gold of the Klamath Mountains of Northern California and South West Oregon
(See California Mining Books)

Mineral Resources of South West Oregon - Unavailable since 1914, this publication includes important insights into dozens of mines that once operated in South West Oregon, including the famous gold fields of Josephine and Jackson Counties, as well as the Coal Mines of Coos County. Included are not only geological details on early mines throughout South West Oregon, but also insights into their history, production and locations. **8.5" X 11", 154 ppgs. Retail Price: $11.99**

Chromite Mining in The Klamath Mountains of California and Oregon
(See California Mining Books)

Southern Oregon Mineral Wealth - Unavailable since 1904, this rare publication provides a unique snapshot into the mines that were operating in the area at the time. Included are not only geological details on early mines throughout South West Oregon, but also insights into their history, production and locations. Some of the mining areas include Grave Creek, Greenback, Wolf Creek, Jump Off Joe Creek, Granite Hill, Galice, Mount Reuben, Gold Hill, Galls Creek, Kane Creek, Sardine Creek, Birdseye Creek, Evans Creek, Foots Creek, Jacksonville, Ashland, the Applegate River, Waldo, Kerby and the Illinois River, Althouse and Sucker Creek, as well as insights into local copper mining and other topics. **8.5" X 11", 64 ppgs. Retail Price: $8.99**

Geology and Ore Deposits of the Takilma and Waldo Mining Districts - Unavailable since the 1933, this publication was originally compiled by the United States Geological Survey and includes details on gold and copper mining in the Takilma and Waldo Districts of Josephine County, Oregon. The Waldo and Takilma mining districts are most notable for the fact that the earliest large scale mining of placer gold and copper in Oregon took place in these two areas. Included in this report are details about some of the earliest large gold mines in the state such as the Llano de Oro, High Gravel, Cameron, Platerica, Deep Gravel and others, as well as copper mines such as the famous Queen of Bronze mine, the Waldo, Lily and Cowboy mines. In addition to geological examinations, insights are also provided into the production, day to day operations and early histories of these mines, as well as calculations of known mineral reserves in the area. This volume also includes six maps and 20 original illustrations. **8.5" X 11", 74 ppgs. Retail Price: $9.99**

Gold Mines of Oregon - Oregon mining historian Kerby Jackson introduces us to a classic work on Oregon's mining history in this important re-issue of Bulletin 61, otherwise known as "Gold and Silver In Oregon". Unavailable since 1968, this important publication was originally compiled by geologists Howard C. Brooks and Len Ramp of the Oregon Department of Geology and Mineral Industries and includes detailed descriptions, histories and the geology of over 450 gold mines Oregon. Included are notes on the history, geology and gold production statistics of all the major mining areas in Oregon including the Klamath Mountains, the Blue Mountains and the North Cascades. While gold is where you find it, as every miner knows, the path to success is to prospect for gold where it was previously found. 8.5" X 11", 344 ppgs. Retail Price: $24.99

Mines and Mineral Resources of Curry County Oregon - Originally published in 1916, this important publication on Oregon Mining has not been available for nearly a century. Included are rare insights into the history, production and locations of dozens of gold mines in Curry County, Oregon, as well as detailed information on important Oregon mining districts in that area such as those at Agness, Bald Face Creek, Mule Creek, Boulder Creek, China Diggings, Collier Creek, Elk River, Gold Beach, Rock Creek, Sixes River and elsewhere. Particular attention is especially paid to the famous beach gold deposits of this portion of the Oregon Coast. 8.5" X 11", 140 ppgs. Retail Price: $11.99

Chromite Mining in South West Oregon - Originally published in 1961, this important publication on Oregon Mining has not been available for nearly a century. Included are rare insights into the history, production and locations of nearly 300 chromite mines in South Western Oregon. 8.5" X 11", 184 ppgs. Retail Price: $14.99

Mineral Resources of Douglas County Oregon - Originally published in 1972, this important publication on Oregon Mining has not been available for nearly forty years. Included are rare insights into the geology, history, production and locations of numerous gold mines and other mining properties in Douglas County, Oregon. 8.5" X 11", 124 ppgs. Retail Price: $11.99

Mineral Resources of Coos County Oregon - Originally published in 1972, this important publication on Oregon Mining has not been available for nearly forty years. Included are rare insights into the geology, history, production and locations of numerous gold mines and other mining properties in Coos County, Oregon. 8.5" X 11", 100 ppgs. Retail Price: $11.99

Mineral Resources of Lane County Oregon - Originally published in 1938, this important publication on Oregon Mining has not been available for nearly seventy five years. Included are extremely rare insights into the geology and mines of Lane County, Oregon, in particular in the Bohemia, Blue River, Oakridge, Black Butte and Winberry Mining Districts. 8.5" X 11", 82 ppgs. Retail Price: $9.99

Mineral Resources of the Upper Chetco River of Oregon: Including the Kalmiopsis Wilderness - Originally published in 1975, this important publication on Oregon Mining has not been available for nearly forty years. Withdrawn under the 1872 Mining Act since 1984, real insight into the minerals resources and mines of the Upper Chetco River has long been unavailable due to the remoteness of the area. Despite this, the decades of battle between property owners and environmental extremists over the last private mining inholding in the area has continued to pique the interest of those interested in mining and other forms of natural resource use. Gold mining began in the area in the 1850's and has a rich history in this geographic area, even if the facts surrounding it are little known. Included are twenty two rare photographs, as well as insights into the Becca and Morning Mine, the Emmly Mine (also known as Emily Camp), the Frazier Mine, the Golden Dream or Higgins Mine, Hustis Mine, Peck Mine and others. 8.5" X 11", 64 ppgs. Retail Price: $8.99

Gold Dredging in Oregon - Originally published in 1939, this important publication on Oregon Mining has not been available for nearly seventy five years. Included are extremely rare insights into the history and day to day operations of the dragline and bucketline gold dredges that once worked the placer gold fields of South West and North East Oregon in decades gone by. Also included are details into the areas that were worked by gold dredges in Josephine, Jackson, Baker and Grant counties, as well as the economic factors that impacted this mining method. This volume also offers a unique look into the values of river bottom land in relation to both farming and mining, in how farm lands were mined, re-soiled and reclamated after the dredges worked them. Featured are hard to find maps of the gold dredge fields, as well as rare photographs from a bygone era. 8.5" X 11", 86 ppgs. Retail Price: $8.99

Quick Silver Mining in Oregon - Originally published in 1963, this important publication on Oregon Mining has not been available for over fifty years. This publication includes details into the history and production of Elemental Mercury or Quicksilver in the State of Oregon. 8.5" X 11", 238 ppgs. Retail Price: $15.99

Mines of the Greenhorn Mining District of Grant County Oregon - Originally published in 1948, this important publication on Oregon Mining has not been available for over sixty five years. In this publication are rare insights into the mines of the famous Greenhorn Mining District of Grant County, Oregon, especially the famous Morning Mine. Also included are details on the Tempest, Tiger, Bi-Metallic, Windsor, Psyche, Big Johnny, Snow Creek, Banzette and Paramount Mines, as well as prospects in the vicinities in the famous mining areas of Mormon Basin, Vinegar Basin and Desolation Creek. Included are hard to find mine maps and dozens of rare photographs from the bygone era of Grant County's rich mining history. 8.5" X 11", 72 ppgs. Retail Price: $9.99

Geology of the Wallowa Mountains of Oregon: Part I (Volume 1) - Originally published in 1938, this important publication on Oregon Mining has not been available for nearly seventy five years. Included are details on the geology of this unique portion of North Eastern Oregon. This is the first part of a two book series on the area. Accompanying the text are rare photographs and historic maps. **8.5" X 11", 92 ppgs. Retail Price: $9.99**

Geology of the Wallowa Mountains of Oregon: Part II (Volume 2) - Originally published in 1938, this important publication on Oregon Mining has not been available for nearly seventy five years. Included are details on the geology of this unique portion of North Eastern Oregon. This is the first part of a two book series on the area. Accompanying the text are rare photographs and historic maps. **8.5" X 11", 94 ppgs. Retail Price: $9.99**

Field Identification of Minerals For Oregon Prospectors - Originally published in 1940, this important publication on Oregon Mining has not been available for nearly seventy five years. Included in this volume is an easy system for testing and identifying a wide range of minerals that might be found by prospectors, geologists and rockhounds in the State of Oregon, as well as in other locales. Topics include how to put together your own field testing kit and how to conduct rudimentary tests in the field. This volume is written in a clear and concise way to make it useful even for beginners. **8.5" X 11", 158 ppgs. Retail Price: $14.99**

The Bohemia Mining District of Oregon - Originally published in 1900, this important publication on Oregon Mining has not been available for over a century. Included in this volume are important insights into the famous Bohemia Mining District of Oregon, including the histories and locations of important gold mines in the area such as the Ophir Mine, Clarence, Acturas, Peek-a-boo, White Swan, Combination Mine, the Musick Mine, The California, White Ghost, The Mystery, Wall Street, Vesuvius, Story, Lizzie Bullock, Delta, Elsie Dora, Golden Slipper, Broadway, Champion Mine, Knott, Noonday, Helena, White Wings, Riverside and others. Also included are notes on the nearby Blue River Mining District. **8.5" X 11", 58 ppgs. Retail Price: $9.99**

The Gold Fields of Eastern Oregon - Unavailable since 1900, this publication was originally compiled by the Baker City Chamber of Commerce Offering important insights into the gold mining history of Eastern Oregon, "The Gold Fields of Eastern Oregon" sheds a rare light on many of the gold mines that were operating at the turn of the 19th Century in Baker County and Grant County in North Eastern Oregon. Some of the areas featured include the Cable Cove District, Baisely-Elhorn, Granite, Red Boy, Bonanza, Susanville, Sparta, Virtue, Vaughn, Sumpter, Burnt River, Rye Valley and other mining districts. Included is basic information on not only many gold mines that are well known to those interested in Eastern Oregon mining history, but also many mines and prospects which have been mostly lost to the passage of time. Accompanying are numerous rare photos **8.5" X 11", 78 ppgs. Retail Price: $10.99**

Gold Mining in Eastern Oregon - Originally published in 1938, this important publication on Oregon Mining has not been available for over a century. Included in this volume are important insights into the famous mining districts of Eastern Oregon during the late 1930's. Particular attention is given to those gold mines with milling and concentrating facilities in the Greenhorn, Red Boy, Alamo, Bonanza, Granite, Cable Cove, Cracker Creek, Virtue, Keating, Medical Springs, Sanger, Sparta, Chicken Creek, Mormon Basin, Connor Creek, Cornucopia and the Bull Run Mining Districts. Some of the mines featured include the Ben Harrison, North Pole-Columbia, Highland Maxwell, Baisley-Elkhorn, White Swan, Balm Creek, Twin Baby, Gem of Sparta, New Deal, Gleason, Gifford-Johnson, Cornucopia, Record, Bull Run, Orion and others. Of particular interest are the mill flow sheets and descriptions of milling operations of these mines. **8.5" X 11", 68 ppgs. Retail Price: $8.99**

The Gold Belt of the Blue Mountains of Oregon - Originally published in 1901, this important publication on Oregon Mining has not been available for over a century. Included in this volume are rare insights into the gold deposits of the Blue Mountains of North East Oregon, including the history of their early discovery and early production. Extensive details are offered on this important mining area's mineralogy and economic geology, as well as insights into nearby gold placers, silver deposits and copper deposits. Featured are the Elkhorn and Rock Creek mining districts, the Pocahontas district, Auburn and Minersville districts, Sumpter and Cracker Creek, Cable Cove, the Camp Carson district, Granite, Alamo, Greenhorn, Robinsonville, the Upper Burnt River Valley and Bonanza districts, Susanville, Quartzburg, Canyon Creek, Virtue, the Copper Butte district, the North Powder River, Sparta, Eagle Creek, Cornucopia, Pine Creek, Lower Powder River, the Upper Snake River Canyon, Rye Valley, Lower Burnt River Valley, Mormon Basin, the Malheur and Clarks Creek districts, Sutton Creek and others. Of particular interest are important details on numerous gold mines and prospects in these mining districts, including their locations, histories, geology and other important information, as well as information on silver, copper and fire opal deposits. **8.5" X 11", 250 ppgs. Retail Price: $24.99**

Mining in the Cascades Range of Oregon - Originally published in 1938, this important publication on Oregon Mining has not been available for over seventy five years. Included in this volume are rare insights into the gold mines and other types of metal mines in the Cascades Mountain Range of Oregon. Some of the important mining areas covered include the famous Bohemia Mining District, the North Santiam Mining District, Quartzville Mining District, Blue River Mining District, Fall Creek Mining District, Oakridge District, Zinc District, Buzzard-Al Sarena District, Grand Cove, Climax District and Barron Mining District. Of particular interest are important details on over 100 mines and prospects in these mining districts, including their locations, histories, geology and other important information. 8.5" X 11", 170 ppgs. Retail Price: $14.99

Beach Gold Placers of the Oregon Coast - Originally published in 1934, this important publication on Oregon Mining has not been available for over 80 years. Included in this volume are rare insights into the beach gold deposits of the State of Oregon, including their locations, occurance, composition and geology. Of particular interest is information on placer platinum in Oregon's rich beach deposits. Also included are the locations and other information on some famous Oregon beach mines, including the Pioneer, Eagle, Chickamin, Iowa and beach placer mines north of the mouth of the Rogue River. 8.5" X 11", 60 ppgs. Retail Price: $8.99

Mineralogical Composition of the Sands of the Oregon Coast: From Coos Bay to the Columbia - Published in 1945, he text features hard to find information on the composition of the gold bearing black sands of the South West Oregon Coast, offering a unique insight to prospectors in search of Oregon's legendary beach gold. 104 ppgs, $9.99

Manganese Mining in Oregon - First released in 1942 and now out of print, this special reprint edition of "Manganese in Oregon" was originally published by the Oregon Department of Geology and Mineral Industries. The text features hard to find information on the mining of Manganese in Oregon, including details and maps of Oregon manganese mines and prospects. 108 ppgs, 9.99

Medford Oregon As A Mining Center - Written in 1912, this hard to find publication includes valuable insights into the mining history of South West Oregon. This small book contains interesting information on the gold, copper and mining industry in Southern Oregon as it existed just prior to World War One, shedding light on some of the important mines in the area. Included are rare photographs and vintage advertising of the day. 80 ppgs, 9.99

Mineral Resources of Curry County Oregon - First released in 1977 and now out of print, this special reprint edition of "Geology, Mineral Resources and Rock Materials of Curry County, Oregon" was originally published in cooperation of Curry County, Oregon and the Oregon Department of Geology and Mineral Industries. The text features hard to find information on not only the mining of gold and other metals in Curry County, but also aggregate mining in the area. 102 ppgs, 11.99

Origin of the Gold Bearing Black Sands of the Coast of South West Oregon - First released in 1943 and now out of print, this special reprint edition of "The Origin of the Black Sands of the South West Oregon Coast" was originally published by the Oregon Department of Geology and Mineral Industries. The text features hard to find information on the origin of the gold bearing black sands of the South West Oregon Coast, offering a unique insight to prospectors in search of Oregon's legendary beach gold. 52 ppgs, 8.99

South West Oregon Mining - Leading mining historian Kerby Jackson introduces us to six classic small mining publications on the Gold Mining Industry in Southern Oregon. This small book consists of a compilation of USGS J.S. Diller's "Mines of the Riddles Quadrangle", "The Rogue River Valley Coal Fields" and "Mineral Resources of the Grants Pass Quadrangle", the Grants Pass Commercial Club's rare publication "Mining in Josephine County, Oregon" and the USGS publication "The Distribution of Placer Gold in the Sixes River, South West Oregon". Also included is F.W. Libbey's legendary article on the Southern Oregon Mining Industry, "Lest We Forget", which appeared in the publication of the Oregon State Department of Geology and Mineral Industries in the early 1960's. This compilation offers a unique perspective on mining in South West Oregon and includes considerable information on mines in Josephine, Jackson and Coos Counties. 142 ppgs, 14.99

Geology and Mineral Resources of the Gasquet Quadrangle of California-Oregon - First published in 1953, it has been unavailable for over a century and sheds important light on the geological features and mineral resources of this portion of Northern California and Southern Oregon. 80 ppgs, 9.99

Idaho Mining Books

Gold in Idaho - Unavailable since the 1940's, this publication was originally compiled by the Idaho Bureau of Mines and includes details on gold mining in Idaho. Included is not only raw data on gold production in Idaho, but also valuable insight into where gold may be found in Idaho, as well as practical information on the gold bearing rocks and other geological features that will assist those looking for placer and lode gold in the State of Idaho. This volume also includes thirteen gold maps that greatly enhance the practical usability of the information contained in this small book detailing where to find gold in Idaho. **8.5" X 11", 72 ppgs. Retail Price: $9.99**

Geology of the Couer D'Alene Mining District of Idaho - Unavailable since 1961, this publication was originally compiled by the Idaho Bureau of Mines and Geology and includes details on the mining of gold, silver and other minerals in the famous Coeur D'Alene Mining District in Northern Idaho. Included are details on the early history of the Coeur D'Alene Mining District, local tectonic settings, ore deposit features, information on the mineral belts of the Osburn Fault, as well as detailed information on the famous Bunker Hill Mine, the Dayrock Mine, Galena Mine, Lucky Friday Mine and the infamous Sunshine Mine. This volume also includes sixteen hard to find maps. **8.5" X 11", 70 ppgs. Retail Price: $9.99**

The Gold Camps and Silver Cities of Idaho - Originally published in 1963, this important publication on Idaho Mining has not been available for nearly fifty years. Included are rare insights into the history of Idaho's Gold Rush, as well as the mad craze for silver in the Idaho Panhandle. Documented in fine detail are the early mining excitements at Boise Basin, at South Boise, in the Owyhees, at Deadwood, Long Valley, Stanley Basin and Robinson Bar, at Atlanta, on the famous Boise River, Volcano, Little Smokey, Banner, Boise Ridge, Hailey, Leesburg, Lemhi, Pearl, at South Mountain, Shoup and Ulysses, Yellow Jacket and Loon Creek. The story follows with the appearance of Chinese miners at the new mining camps on the Snake River, Black Pine, Yankee Fork, Bay Horse, Clayton, Heath, Seven Devils, Gibbonsville, Vienna and Sawtooth City. Also included are special sections on the Idaho Lead and Silver mines of the late 1800's, as well as the mining discoveries of the early 1900's that paved the way for Idaho's modern mining and mineral industry. Lavishly illustrated with rare historic photos, this volume provides a one of a kind documentary into Idaho's mining history that is sure to be enjoyed by not only modern miners and prospectors who still scour the hills in search of nature's treasures, but also those enjoy history and tromping through overgrown ghost towns and long abandoned mining camps. **8.5" X 11", 186 ppgs. Retail Price: $14.99**

Ore Deposits and Mining in North Western Custer County Idaho - Unavailable since 1913, this important publication was originally published by the Us Department of the Interior and has been unavailable for a century. Included are fine details on the geology, geography, gold placers and gold and silver bearing quartz veins of the mining region of North West Custer County, Idaho. Of particular interest is a rare look at the mines and prospects of the region, including those such as the Ramshorn Mine, SkyLark, Riverview, Excelsior, Beardsley, Pacific, Hoosier, Silver Brick, Forest Rose and dozens of others in the Bay Horse Mining District. Also covered are the mines of the Yankee Fork District such as the Lucky Boy, Badger, Black, Enterprise, Charles Dickens, Morrison, Golden Sunbeam, Montana, Golden Gate and others, as well as those in the Loon Mining District. **8.5" X 11", 126 ppgs. Retail Price: $12.99**

Gold Rush To Idaho - Unavailable since 1963, this important publication was originally published by the Idaho Bureau of Mines and has been unavailable for 50 years. "Gold Rush To Idaho" revisits the earliest years of the discovery of gold in Idaho Territory and introduces us to the conditions that the pioneer gold seekers met when they blazed a trail through the wilderness of Idaho's mountains and discovered the precious yellow metal at Oro Fino and Pierce. Subsequent rushes followed at places like Elk City, Newsome, Clearwater Station, Florence, Warrens and elsewhere. Of particular interest is a rare look at the hardships that the first miners in Idaho met with during their day to day existences and their attempts to bring law and order to their mining camps. **8.5" X 11", 88 ppgs. Retail Price: $9.99**

The Geology and Mines of Northern Idaho and North Western Montana - Unavailable since 1909, this important publication was originally published by the Us Department of the Interior and has been unavailable for a century. Included are fine details on the geology and geography of the mining regions of Northern Idaho and North Western Montana. Of particular interest is a rare look at the mines and prospects of the region, including those in the Pine Creek Mining District, Lake Pend Oreille district, Troy Mining District, Sylvanite District, Cabinet Mining District, Prospect Mining District and the Missoula Valley. Some of the mines featured include the Iron Mountain, Silver Butte, Snowshoe, Grouse Mountain Mine and others. **8.5" X 11", 142 ppgs. Retail Price: $12.99**

Mining in the Alturas Quadrangle of Blaine County Idaho - Unavailable since 1922, this important publication was originally published by the Idaho Bureau of Mines and has been unavailable for ninety years. Topics include the geology, rock formations and the formation of ore deposits in this important mining area of Idaho. Of particular focus is information on the local geology, quartz veins and ore deposits of this portion of Idaho. Included are hard to find details, including the descriptions and locations of numerous gold and silver mines in the area including the Silver King, Pilgrim, Columbia, Lone Jack, Sunbeam, Pride of the West, Lucky Boy, Scotia, Atlanta, Beaver-Bidwell and others mines and prospects. **8.5" X 11", 56 ppgs. Retail Price: $8.99**

Mining in Lemhi County Idaho - Originally published in 1913, this important book on Idaho Mining has not been available to miners for over a century. Included are rare insights into hundreds of gold, silver, copper and other mines in this famous Idaho mining area. Details include the locations, geology, history, production and other facts of the mines of this region, not only gold and silver hardrock mines, but also gold placer mines, lead-silver deposits, copper mines, cobalt-nickel deposits, tungsten and tin mines . It is lavishly illustrated with hard to find photos of the period and rare mining maps. Some of the vicinities featured include the Nicholia Mining District, Spring Mountain District, Texas District, Blue Wing District, Junction District, McDevitt District, Pratt Creek, Eldorado District, Kirtley Creek, Carmen Creek, Gibbonsville, Indian Creek, Mineral Hill District, Mackinaw, Eureka District, Blackbird District, YellowJacket District, Gravel Range District, Junction District, Parker Mountain and other mining districts. 8.5" X 11", 226 ppgs. Retail Price: $19.99

Mining in Shoshone County Idaho - First published in 1923, it has been unavailable for over a century and sheds important light on the mining history of Shoshone County, Idaho. Some of the topics include the history of mining in Shoshone County, a look at the local geology and ore characteristics of lead-silver deposits, zinc deposits, copper, antimony, gold and other minerals. Also included are insights into the history, production, characteristics and locations of numerous mines in the area. 198 ppgs, 15.99

Utah Mining Books

Fluorite in Utah - Unavailable since 1954, this publication was originally compiled by the USGS, State of Utah and U.S. Atomic Energy Commission and details the mining of fluorspar, also known as fluorite in the State of Utah. Included are details on the geology and history of fluorspar (fluorite) mining in Utah, including details on where this unique gem mineral may be found in the State of Utah. 8.5" X 11", 60 ppgs. Retail Price: $8.99

The Gold Hill Mining District of Utah - First published in 1935, it has been unavailable since those days and sheds important light on the mines, history and geology of Utah's Gold Hill Mining District. Included are rare insights into this important mining area, including the locations, histories and details of numerous mines. This volume is well illustrated with geological diagrams, as well as hard to find maps of some of the most important mines in this district. 202 ppgs., 19.99

The Mines, Miners and Minerals of Utah - First published in 1896, it has been unavailable since those days and sheds important light on the early mines and miners of Pioneer Utah, as well as the minerals which they won from the earth by laborious hard physical labor and sheer determination. Included are rare insights into the early mining history of Utah, as well details on hundreds of gold, silver and copper mines. 376 ppgs., 24.99

California Mining Books

The Tertiary Gravels of the Sierra Nevada of California - Mining historian Kerby Jackson introduces us to a classic mining work by Waldemar Lindgren in this important re-issue of The Tertiary Gravels of the Sierra Nevada of California. Unavailable since 1911, this publication includes details on the gold bearing ancient river channels of the famous Sierra Nevada region of California. 8.5" X 11", 282 ppgs. Retail Price: $19.99

The Mother Lode Mining Region of California - Unavailable since 1900, this publication includes details on the gold mines of California's famous Mother Lode gold mining area. Included are details on the geology, history and important gold mines of the region, as well as insights into historic mining methods, mine timbering, mining machinery, mining bell signals and other details on how these mines operated. Also included are insights into the gold mines of the California Mother Lode that were in operation during the first sixty years of California's mining history. 8.5" X 11", 176 ppgs. Retail Price: $14.99

Lode Gold of the Klamath Mountains of Northern California and South West Oregon - Unavailable since 1971, this publication was originally compiled by Preston E. Hotz and includes details on the lode mining districts of Oregon and California's Klamath Mountains. Included are details on the geology, history and important lode mines of the French Gulch, Deadwood, Whiskeytown, Shasta, Redding, Muletown, South Fork, Old Diggings, Dog Creek (Delta), Bully Choop (Indian Creek), Harrison Gulch, Hayfork, Minersville, Trinity Center, Canyon Creek, East Fork, New River, Denny, Liberty (Black Bear), Cecilville, Callahan, Yreka, Fort Jones and Happy Camp mining districts in California, as well as the Ashland, Rogue River, Applegate, Illinois River, Takilma, Greenback, Galice, Silver Peak, Myrtle Creek and Mule Creek districts of South Western Oregon. Also included are insights into the mineralization and other characteristics of this important mining region. 8.5" X 11", 100 ppgs. Retail Price: $10.99

Mines and Mineral Resources of Shasta County, Siskiyou County, Trinity County: California - Unavailable since 1915, this publication was originally compiled by the California State Mining Bureau and includes details on the gold mines of this area of Northern California. Also included are insights into the mineralization and other characteristics of this important mining region, as well as the location of historic gold mines. 8.5" X 11", 204 ppgs. Retail Price: $19.99

Geology of the Yreka Quadrangle, Siskiyou County, California - Unavailable since 1977, this publication was originally compiled by Preston E. Hotz and includes details on the geology of the Yreka Quadrangle of Siskiyou County, California. Also included are insights into the mineralization and other characteristics of this important mining region. 8.5" X 11", 78 ppgs. Retail Price: $7.99

Mines of San Diego and Imperial Counties, California - Originally published in 1914, this important publication on California Mining has not been available for a century. This publication includes important information on the early gold mines of San Diego and Imperial County, which were some of the first gold fields mined in California by early Spanish and Mexican miners before the 49ers came on the scene. Included are not only details on early mining methods in the area, production statistics and geological information, but also the location of the early gold mines that helped make California "The Golden State". Also included are details on the mining of other minerals such as silver, lead, zinc, manganese, tungsten, vanadium, asbestos, barite, borax, cement, clay, dolomite, fluospar, gem stones, graphite, marble, salines, petroleum, stronium, talc and others. 8.5" X 11", 116 ppgs. Retail Price: $12.99

Mines of Sierra County, California - Unavailable since 1920, this publication was originally compiled by the California State Mining Bureau and includes details on the gold mines of Sierra County, California. Also included are insights into the mineralization and other characteristics of this important mining region, as well as the location of historic gold mines. 8.5" X 11", 156 ppgs. Retail Price: $19.99

Mines of Plumas County, California - Unavailable since 1918, this publication was originally compiled by the California State Mining Bureau and includes details on the gold mines of Plumas County, California. Also included are insights into the mineralization and other characteristics of this important mining region, as well as the location of historic gold mines. 8.5" X 11", 200 ppgs. Retail Price: $19.99

Mines of El Dorado, Placer, Sacramento and Yuba Counties, California - Originally published in 1917, this important publication on California Mining has not been available for nearly a century. This publication includes important information on the early gold mines of El Dorado County, Placer County, Sacramento County and Yuba County, which were some of the first gold fields mined by the Forty-Niners during the California Gold Rush. Included are not only details on early mining methods in the area, production statistics and geological information, but also the location of the early gold mines that helped make California "The Golden State". Also included are insights into the early mining of chrome, copper and other minerals in this important mining area. 8.5" X 11", 204 ppgs. Retail Price: $19.99

Mines of Los Angeles, Orange and Riverside Counties, California - Originally published in 1917, this important publication on California Mining has not been available for nearly a century. This publication includes important information on the early gold mines of Los Angeles County, Orange County and Riverside County, which were some of the first gold fields mined in California by early Spanish and Mexican miners before the 49ers came on the scene. Included are not only details on early mining methods in the area, production statistics and geological information, but also the location of the early gold mines that helped make California "The Golden State". 8.5" X 11", 146 ppgs. Retail Price: $12.99

Mines of San Bernadino and Tulare Counties, California - Originally published in 1917, this important publication on California Mining has not been available for nearly a century. This publication includes important information on the early gold mines of San Bernadino and Tulare County, which were some of the first gold fields mined in California by early Spanish and Mexican miners before the 49ers came on the scene. Included are not only details on early mining methods in the area, production statistics and geological information, but also the location of the early gold mines that helped make California "The Golden State". Also included are details on the mining of other minerals such as copper, iron, lead, zinc, manganese, tungsten, vanadium, asbestos, barite, borax, cement, clay, dolomite, fluospar, gem stones, graphite, marble, salines, petroleum, stronium, talc and others. 8.5" X 11", 200 ppgs. Retail Price: $19.99

Chromite Mining in The Klamath Mountains of California and Oregon - Unavailable since 1919, this publication was originally compiled by J.S. Diller of the United States Department of Geological Survey and includes details on the chromite mines of this area of Northern California and Southern Oregon. Also included are insights into the mineralization and other characteristics of this important mining region, as well as the location of historic mines. Also included are insights into chromite mining in Eastern Oregon and Montana. 8.5" X 11", 98 ppgs. Retail Price: $9.99

Mines and Mining in Amador, Calaveras and Tuolumne Counties, California - Unavailable since 1915, this publication was originally compiled by William Tucker and includes details on the mines and mineral resources of this important California mining area. Included are details on the geology, history and important gold mines of the region, as well as insights into other local mineral resources such as asbestos, clay, copper, talc, limestone and others. Also included are insights into the mineralization and other characteristics of this important portion of California's Mother Lode mining region. 8.5" X 11", 198 ppgs. Retail Price: $14.99

The Cerro Gordo Mining District of Inyo County California - Unavailable since 1963, this publication was originally compiled by the United States Department of Interior. Included are insights into the mineralization and other characteristics of this important mining region of Southern California. Topics include the mining of gold and silver in this important mining district in Inyo County, California, including details on the history, production and locations of the Cerro Gordo Mine, the Morning Star Mine, Estelle Tunnel, Charles Lease Tunnel, Ignacio, Hart, Crosscut Tunnel, Sunset, Upper Newtown, Newtown, Ella, Perseverance, Newsboy, Belmont and other silver and gold mines in the Cerro Gordo Mining District. This volume also includes important insights into the fossil record, geologic formations, faults and other aspects of economic geology in this California mining district. 8.5" X 11", 104 ppgs. Retail Price: $10.99

Mining in Butte, Lassen, Modoc, Sutter and Tehama Counties of California - Unavailable since 1917, this publication was originally compiled by the United States Department of Interior. Included are insights into the mineralization and other characteristics of this important mining region of California. Topics include the mining of asbestos, chromite, gold, diamonds and manganese in Butte County, the mining of gold and copper in the Hayden Hill and Diamond Mountain mining districts of Lassen County, the mining of coal, salt, copper and gold in the High Grade and Winters mining districts of Modoc County, gold mining in Sutter County and the mining of gold, chromite, manganese and copper in Tehama County. This volume also includes the production records and locations of numerous mines in this important mining region. 8.5" X 11", 114 ppgs. Retail Price: $11.99

Mines of Trinity County California - Originally published in 1965, this important publication on California Mining has not been available for nearly fifty years. This publication includes important information on mines and mining in Trinity County, California, as well insights into the mineralization and geology of this important mining area in Northern California. Included are extensive details on hardrock and placer gold mines and prospects, including charts showing the locations of these historic mines.. 8.5" X 11", 144 ppgs. Retail Price: $12.99

Mines of Kern County California - Originally published in 1962, this important publication on California Mining has not been available for nearly fifty years. This publication includes important information on mines and mining in Kern County, California, as well insights into the mineralization and geology of this important mining area in California. Included are extensive details on hardrock and placer gold mines and prospects, including charts showing the locations of these historic mines. 8.5" X 11", 398 ppgs. Retail Price: $24.99

Mines of Calaveras County California - Originally published in 1962, this important publication on California Mining has not been available for nearly fifty years. This publication includes important information on mines and mining in Calaveras County, California, as well insights into the mineralization and geology of this important mining area in Northern California. Included are extensive details on hardrock and placer gold mines and prospects, including charts showing the locations of these historic mines. 8.5" X 11", 236 ppgs. Retail Price: $19.99

Lode Gold Mining in Grass Valley California - Unavailable since 1940, this publication was originally compiled by the United States Department of Interior. Included are insights into the gold mineralization and other characteristics of this important mining region of Nevada County, California. This volume also includes important insights into the geologic formations, faults and other aspects of economic geology in this California mining district. Of particular interest are the fine details on many hardrock gold mines in the area, including their locations, histories, development and mineralization. Some of the mines featured include the Gold Hill Mine, Massachusetts Hill, Boundary, Peabody, Golden Center, North Star, Omaha, Lone Jack, Homeward Bound, Hartery, Wisconsin, Allison Ranch, Phoenix, Kate Hayes, W.Y.O.D., Empire, Rich Hill, Daisy Hill, Orleans, Sultana, Centennial, Conlin, Ben Franklin, Crown Point and many others. 8.5" X 11", 148 ppgs. Retail Price: $12.99

Lode Mining in the Alleghany District of Sierra County California - Unavailable since 1913, this publication was originally compiled by the United States Department of Interior. Included are insights into the mineralization and other characteristics of this important mining region of Sierra County. Included are details on the history, production and locations of numerous hardrock gold mines in this famous California area, including the Tightner Mine, Minnie D., Osceola, Eldorado, Twenty One, Sherman, Kenton, Oriental, Rainbow, Plumbago, Irelan, Gold Canyon, North Fork, Federal, Kate Hardy and others. This volume also includes important insights into the fossil record, geologic formations, faults and other aspects of economic geology in this California mining district. 8.5" X 11", 48 ppgs. Retail Price: $7.99

Six Months In The Gold Mines During The California Gold Rush - Unavailable since 1850, this important work is a first hand account of one "49'ers" personal experience during the great California Gold Rush, shedding important light on one of the most exciting periods in the history of not only California, but also the world. Compiled from journals written between 1847 and 1849 by E. Gould Buffum, a native of New York, "Six Months In The Gold Mines During The California Gold Rush" offers a rare look into the day to day lives of the people who came to California to work in her gold mines when the state was still a great frontier. 8.5" X 11", 290 ppgs. Retail Price: $19.99

Quartz Mines of the Grass Valley Mining District of California - Unavailable since 1867, this important publication has not been available since those days. This rare publication offers a short dissertation on the early hardrock mines in this important mining district in the California Mother Lode region between the 1850's and 1860's. Also included are hard to find details on the mineralization and locations of these mines, as well as how they were operated in those day. **8.5″ X 11″, 44 ppgs. Retail Price: $8.99**

Gold Rush on the Feather River - First published in 1924, this short publication by G.C. Mansfield sheds important light on the early history of gold mining on the Feather River. Included are rare insights into the first decade of gold mining and the early mining camps of the Feather River during the 1850's. **64 ppgs., 9.99**

The Bodie Mining District of California - First published in 1986, it has been unavailable since those days and sheds important light on this famous mining area. Included are the history, characteristics and locations of numerous old mines around the ghost town of Bodie. **64 ppgs, 8.99**

Geology and Mineral Resources of the Gasquet Quadrangle of California-Oregon - First published in 1953, it has been unavailable for over a century and sheds important light on the geological features and mineral resources of this portion of Northern California and Southern Oregon. **80 ppgs, 9.99**

Alaska Mining Books

Ore Deposits of the Willow Creek Mining District, Alaska - Unavailable since 1954, this hard to find publication includes valuable insights into the Willow Creek Mining District near Hatcher Pass in Alaska. The publication includes insights into the history, geology and locations of the well known mines in the area, including the Gold Cord, Independence, Fern, Mabel, Lonesome, Snowbird, Schroff-O'Neil, High Grade, Marion Twin, Thorpe, Webfoot, Kelly-Willow, Lane, Holland and others. **8.5″ X 11″, 96 ppgs. Retail Price: $9.99**

The Juneau Gold Belt of Alaska - Unavailable since 1906, this hard to find publication includes valuable insights into the gold mines around Juneau, Alaska. The publication includes important details into the history, geology and locations of the well known gold mines and prospects in the area, including those around Windham Bay, Holkham Bay, Port Snettisham, on Grindstone and Rhine Creeks, Gold Creek, Douglas Island, Salmon Creek, Lemon Creek, Nugget Creek, from the Mendenhall River to Berners Bay, McGinnis Creek, Montana Creek, Peterson Creek, Windfall Creek, the Eagle River, Yankee Basin, Yankee Curve, Kowee Creek and elsewhere. Not only are gold placer mines included, but also hardrock gold mines. **8.5″ X 11″, 224 ppgs. Retail Price: $19.99**

Mining in the Jumbo Basin of Alaska - Unavailable since 1953, this hard to find publication includes valuable insights into the mines and geology of the Jumbo Basin. The publication includes important details into the history, geology and locations of the well known gold mines and prospects in the famous Jumbo Basin Mining Region of Alaska. **72 ppgs, 9.99**

The Rampart Placer Gold Region of Alaska - Unavailable since 1906, this hard to find publication includes valuable insights into the placer gold mines of the Rampart Mining Region. The publication includes important details into the history, geology and locations of the well known gold mines and prospects in the famous Rampart Mining Region of Alaska. **78 ppgs, 10.99**

Arizona Mining Books

Mines and Mining in Northern Yuma County Arizona - Originally published in 1911, this important publication on Arizona Mining has not been available for over a hundred years. Included are rare insights into the gold, silver, copper and quicksilver mines of Yuma County, Arizona together with hard to find maps and photographs. Some of the mines and mining districts featured include the Planet Copper Mine, Mineral Hill, the Clara Consolidated Mine, Viati Mine, Copper Basin prospect, Bowman Mine, Quartz King, Billy Mack, Carnation, the Wardwell and Osbourne, Valensuella Copper, the Mariquita, Colonial Mine, the French American, the New York-Plomosa, Guadalupe, Lead Camp, Mudersbach Copper Camp, Yellow Bird, the Arizona Northern (Salome Strike), Bonanza (Harqua Hala), Golden Eagle, Hercules, Socorro and others. **8.5″ X 11″, 144 ppgs. Retail Price: $11.99**

The Aravaipa and Stanley Mining Districts of Graham County Arizona - Originally published in 1925, this important publication on Arizona Mining has not been available for nearly ninety years. Included are rare insights into the gold and silver mines of these two important mining districts, together with hard to find maps. **8.5″ X 11″, 140 ppgs. Retail Price: $11.99**

Gold in the Gold Basin and Lost Basin Mining Districts of Mohave County, Arizona - This volume contains rare insights into the geology and gold mineralization of the Gold Basin and Lost Basin Mining Districts of Mohave County, Arizona that will be of benefit to miners and prospectors. Also included is a significant body of information on the gold mines and prospects of this portion of Arizona. This volume is lavishly illustrated with rare photos and mining maps. 8.5" X 11", 188 ppgs. Retail Price: $19.99

Mines of the Jerome and Bradshaw Mountains of Arizona - This important publication on Arizona Mining has not been available for ninety years. This volume contains rare insights into the geology and ore deposits of the Jerome and Bradshaw Mountains of Arizona that will be of benefit to miners and prospectors who work those areas. Included is a significant body of information on the mines and prospects of the Verde, Black Hills, Cherry Creek, Prescott, Walker, Groom Creek, Hassayampa, Bigbug, Turkey Creek, Agua Fria, Black Canyon, Peck, Tiger, Pine Grove, Bradshaw, Tintop, Humbug and Castle Creek Mining Districts. This volume is lavishly illustrated with rare photos and mining maps. 8.5" X 11", 218 ppgs. Retail Price: $19.99

The Ajo Mining District of Pima County Arizona - This important publication on Arizona Mining has not been available for nearly seventy years. This volume contains rare insights into the geology and mineralization of the Ajo Mining District in Pima County, Arizona and in particular the famous New Cornelia Mine. 8.5" X 11", 126 ppgs. Retail Price: $11.99

Mining in the Santa Rita and Patagonia Mountains of Arizona - Originally published in 1915, this important publication on Arizona Mining has not been available for nearly a century. Included are rare insights into hundreds of gold, silver, copper and other mines in this famous Arizona mining area. Details include the locations, geology, history, production and other facts of the mines of this region. 8.5" X 11", 394 ppgs. Retail Price: $24.99

Mining in the Bisbee Quadrangle of Arizona - Originally published in 1906, this important publication on Arizona Mining has not been available for nearly a century. Included are rare insights into hundreds of gold, silver, copper and other mines in this famous Arizona mining area. Details include the locations, geology, history, production and other facts of the mines of this important mining region. 8.5" X 11", 188 ppgs. Retail Price: $14.99

Placer Gold Mining in Arizona - Unavailable since 1922, this hard to find publication includes valuable insights into the placer gold mines of the Arizona. Originally released as "Placer Gold of Arizona", despite its small size, this publication includes important details into the history, geology and locations of the well known placer gold mines and prospects in the State of Arizona. 48 ppgs, 8.99

Gold and Copper Mining near Payson, Arizona - Written in 1915, this hard to find publication includes valuable insights into the gold and copper mining industry of Arizona. Highlighted here are the gold and copper mines near Payson, Arizona. 68 ppgs, 8.99

Lode Gold Mining in Arizona - Unavailable since 1934, this hard to find publication, originally released as "Arizona Lode Gold Mines and Gold Mining" includes valuable insights into the gold mining industry of Arizona. Included are valuable insights into over 150 hardrock gold mines in over 30 different mining districts in Arizona. 278 ppgs, 21.99

Mining in the Dragoon Quadrangle of Cochise County, Arizona - Unavailable since 1964, this hard to find publication includes valuable insights into the mines of the Dragoon Quadrangle Mining Region. The publication includes important details into the history, geology and locations of the well known mines and prospects in this famous mining region of Arizona. 224 ppgs., 19.99

Directory of Operating Mines in Arizona in 1915 - Unavailable since 1916, this hard to find publication includes valuable insights into the mines of Arizona. This small publication includes a complete list of the mines that were operating in the State of Arizona during 1915 and includes details such as general location, owners and some basic facts about each mining operation.52 ppgs. 8.99

Arizona Ore Deposits - Unavailable since 1938, this hard to find publication includes valuable insights into some ore deposits of Arizona. Included are valuable insights into the formation and characteristics of valuable ore deposits in the Jerome, Miami, Inspiration, Clifton, Morenci, Ray, Ajo, Eureka, Tombstone and Magma mining districts. Included are details into some of the major gold, silver and copper mines of these important Arizona mining areas. 160 ppgs, 14.99

Montana Mining Books

A History of Butte Montana: The World's Greatest Mining Camp - First published in 1900 by H.C. Freeman, this important publication sheds a bright light on one of the most important mining areas in the history of The West. Together with his insights, as well as rare photographs of the periods, Harry Freeman describes Butte and its vicinity from its early beginnings, right up to its flush years when copper flowed from its mines like a river. At the time of publication, Butte, Montana was known worldwide as "The Richest Mining Spot On Earth" and produced not only vast amounts of copper, but also silver, gold and other metals from its mines. Freeman illustrates, with great detail, the most important mines in the vicinity of Butte, providing rare details on their owners, their history and most importantly, how the mines operated and how their treasures were extracted. Of particular interest are the dozens of rare photographs that depict mines such as the famous Anaconda, the Silver Bow, the Smoke House, Moose, Paulin, Buffalo, Little Minah, the Mountain Consolidated, West Greyrock, Cora, the Green Mountain, Diamond, Bell, Parnell, the Neversweat, Nipper, Original and many others. **8.5" X 11", 142 ppgs. Retail Price: $12.99**

The Butte Mining District of Montana - This important publication on Montana Mining has not been available for over a century. Included are rare insights into the gold, copper and silver mines of Butte, Montana together with hard to find maps and photographs. Some of the topics include the early history of gold, silver and copper mining in the Butte area, insight into the geology of its mining areas, the local distribution of gold, silver and copper ores, as well their composition and how to identify them. Also included are detailed facts about the mines in the Butte Mining District, including the famous Anaconda Mine, Gagnon, Parrot, Blue Vein, Moscow, Poulin, Stella, Buffalo, Green Mountain, Wake Up Jim, the Diamond-Bell Group, Mountain Consolidated, East Greyrock, West Greyrock, Snowball, Corra, Speculator, Adirondack, Miners Union, the Jessie-Edith May Group, Otisco, Iduna, Colorado, Lizzie, Cambers, Anderson, Hesperus, Preferencia and dozens of others. **8.5" X 11", 298 ppgs. Retail Price: $24.99**

Mines of the Helena Mining Region of Montana - This important publication on Montana Mining has not been available for over a century. Included are rare insights into the gold, copper and silver mines of the vicinity of Helena, Montana, including the Marysville Mining District, Elliston Mining District, Rimini Mining District, Helena Mining District, Clancy Mining District, Wickes Mining District, Boulder and Basin Mining Districts and the Elkhorn Mining District. Some of the topics include the early history of gold, silver and copper mining in the Helena area, insight into the geology of its mining areas, the local distribution of gold, silver and copper ores, as well their composition and how to identify them. Also included are detailed facts, history, geology and locations of over one hundred gold, silver and copper mines in the area . **8.5" X 11", 162 ppgs, Retail Price: $14.99**

Mines and Geology of the Garnet Range of Montana - This important publication on Montana Mining has not been available for over a century. Included are rare insights into the gold, copper and silver mines of the vicinity of this important mining area of Montana. Some of the topics include the early history of gold, silver and copper mining in the Garnet Mountains, insight into the geology of its mining areas, the local distribution of gold, silver and copper ores, as well their composition and how to identify them. Also included are detailed facts, history, geology and locations of numerous gold, silver and copper mines in the area . **8.5" X 11", 100 ppgs, Retail Price: $11.99**

Mines and Geology of the Philipsburg Quadrangle of Montana - This important publication on Montana Mining has not been available for over a century. Included are rare insights into the gold, copper and silver mines of the vicinity of this important mining area of Montana. Some of the topics include the early history of gold, silver and copper mining in the Philipsburg Quadrangle, insight into the geology of its mining areas, the local distribution of gold, silver and copper ores, as well their composition and how to identify them. Also included are detailed facts, history, geology and locations of over one hundred gold, silver and copper mines in the area **8.5" X 11", 290 ppgs, Retail Price: $24.99**

Geology of the Marysville Mining District of Montana - Included are rare insights into the mining geology of the Marysville Mining District. Some of the topics include the early history of gold, silver and copper mining in the area, insight into the geology of its mining areas, the local distribution of gold, silver and copper ores, as well their composition and how to identify them. Also included are detailed facts, history, geology and locations of gold, silver and copper mines in the area **8.5" X 11", 198 ppgs, Retail Price: $19.99**

The Geology and Mines of Northern Idaho and North Western Montana - See listing under Idaho.

The History of Gold Dredging in Montana - Unavailable since 1916, this important publication was originally published by the Us Bureau of Mines and has been unavailable for a century. A century and more ago, giant dredging machines dug in Montana's rivers and creeks in search of illusive golden riches. First appearing in California in the 1850's, gold dredges finally reached their peak of development in Siberia and New Zealand before becoming popular again in the United States. This book offers a unique historical perspective on the gold dredges that once operated in Montana. This book on Montana mining history is lavishly illustrated with dozens of rare historic photos gold dredges that once operated in Montana, as well as hard to locate plans on how these dredges were designed. 120 ppgs., 11.99

Nevada Mining Books

The Bull Frog Mining District of Nevada - Unavailable since 1910, this publication was originally compiled by the United States Department of Interior. This volume also includes important insights into the geologic formations, faults and other aspects of economic geology in this Nevada mining district. Of particular interest are the fine details on many mines in the area, including their locations, histories, development and mineralization. Some of the mines featured include the National Bank Mine, Providence, Gibraltor, Tramps, Denver, Original Bullfrog, Gold Bar, Mayflower, Homestake-King and other mines and prospects. **8.5" X 11", 152 ppgs, Retail Price: $14.99**

History of the Comstock Lode - Unavailable since 1876, this publication was originally released by John Wiley & Sons. This volume also includes important insights into the famous Comstock Lode of Nevada that represented the first major silver discovery in the United States. During its spectacular run, the Comstock produced over 192 million ounces of silver and 8.2 million ounces of gold. Not only did the Comstock result in one of the largest mining rushes in history and yield immense fortunes for its owners, but it made important contributions to the development of the State of Nevada, as well as neighboring California. Included here are important details on not only the early development and history of the Comstock, but also rare early insight into its mines, ore and its geology. **8.5" X 11", 244 ppgs, Retail Price: $19.99**

The Pioche Mining District of Nevada - First published in 1932, it has been unavailable for over a century and sheds important light on the mining history of Nevada. Some of the topics include the history of mining in this district, as well as the characteristics of its mineral and ore deposits. Also included are insights into the history, production, characteristics and locations of numerous mines in the area. Some of the mines include the Combined Metals, Pioche, Ely Valley, No. 10, Poorman, Wide Awake, Alps, Prince, Virginia Louise, Half Moon, Abe Lincoln, Fairview, Bristol Silver, National, Vesuvius, Inman, Tempest, Hillside, Jackrabbit, Lucky Star, Fortuna, Mendha, Manhattan, Hamburg, Comet, Lyndon and others. **108 ppgs 10.99**

The Yerington Mining District of Nevada - First published in 1932, it has been unavailable for over a century and sheds important light on the mining history of Nevada. Some of the topics include the history of mining in this district, as well as the characteristics of its mineral and ore deposits. Also included are insights into the history, production, characteristics and locations of numerous mines in the area. Some of the mines include the Bluestone, Mason Valley, Malachite, McConnell, Greenwood, Western Nevada, Ludwig, Douglas Hill, Casting Copper, Montana-Yerington, Empire, Jim Beatty, Terry and McFarland, Blue Jay and others. **92 ppgs, 10.99**

The Genesis of the Ores of Tonopah Nevada - Unavailable since 1918, this hard to find publication includes valuable insights into the gold mines around Tonopah, Nevada. The publication includes important details into the geology of mines in the Tonopah Mining District of Nevada. **90 ppgs, 10.99**

Mining Camps of Elko, Lander and Eureka Counties Nevada - Unavailable since 1910, this hard to find publication includes valuable insights into the mining camps of Elko, Lander and Eureka Counties, Nevada. The publication includes important details into the history of mines and mining in these three Nevada counties. **154 ppgs, 12.99**

Ore Deposits of the Bullfrog Quadrangle - Unavailable since 1964 and released as "Geology of Bullfrog Quadrangle and Ore Deposits Related to Bullfrog Hills Caldera, Nye County, Nevada and Inyo County, California". The publication includes important details into the geology of mines in the Bullfrog Quadrangle of Nye County, Nevada and Inyo County, California. **52 ppgs, 9.99**

Mining in Eureka County Nevada - Unavailable since 1879, this hard to find publication includes valuable insights into the early mining history off Eureka County, Nevada. The publication includes important details into the early history of the mines of Eureka County, as well as their development, production and how their ores were treated. Also included are details on the 1872 Mining Act, as well as the local rules, regulations and customs of the miners in Eureka County. **134 ppgs, 12.99**

Colorado Mining Books

Ores of The Leadville Mining District - Unavailable since 1926, this publication was originally compiled by the United States Department of Interior. This volume also includes important insights into the ores and mineralization of the Leadville Mining District in Colorado. Topics include historic ore prospecting methods, local geology, insights into ore veins and stockworks, the local trend and distribution of ore channels, reverse faults, shattered rock above replacement ore bodies, mineral enrichment in oxidized and sulphide zones and more. **8.5" X 11", 66 ppgs, Retail Price: $8.99**

Mining in Colorado - Unavailable since 1926, this publication was originally compiled by the United States Department of Interior. This volume also includes important insights into the mining history of Colorado from its early beginnings in the 1850's right up to the mid 1920's. Not only is Colorado's gold mining heritage included, but also its silver, copper, lead and zinc mining industry. Each mining area is treated separately, detailing the development of Colorado's mines on a county by county basis. **8.5" X 11", 284 ppgs, Retail Price: $19.99**

Gold Mining in Gilpin County Colorado - Unavailable since 1876, this publication was originally compiled by the Register Steam Printing House of Central City, Colorado. A rare glimpse at the gold mining history and early mines of Gilpin County, Colorado from their first discovery in the 1850's up to the "flush years" of the mid 1870's. Of particular interest is the history of the discovery of gold in Gilpin County and details about the men who made those first strikes. Special focus is given to the early gold mines and first mining districts of the area, many of which are not detailed in other books on Colorado's gold mining history. **8.5" X 11", 156 ppgs, Retail Price: $12.99**

Mining in the Gold Brick Mining District of Colorado - Important insights into the history of the Gold Brick Mining District, as well as its local geography and economic geology. Also included are the histories and locations of historic mines in this important Colorado Mining District, including the Cortland, Carter, Raymond, Gold Links, Sacramento, Bassick, Sandy Hook, Chronicle, Grand Prize, Chloride, Granite Mountain, Lucille, Gray Mountain, Hilltop, Maggie Mitchell, Silver Islet, Revenue, Roosevelt, Carbonate King and others. In addition to hardrock mining, are also included are details on gold placer mining in this portion of Colorado. **8.5" X 11", 140 ppgs, Retail Price: $12.99**

Ore Deposits of the London Fault of Colorado - First published in 1941, it has been unavailable since those days and sheds important light on the mines and mineral deposits of the London Fault in Central Colorado's Alma Mining District. This publication sheds important light on the gold veins and lead-silver deposits of the Alma Mining District. Included are geologic details on the London Mine, American Mine, Havigorst Tunnel, Ophir Mine, Mosher Tunnel, London-Butte Mine, Venture Shaft, Hard-To-Beat Mine, Oliver Twist Tunnel, Sacramento Mine, Mudsill Mine, Sherwood Mine, Wagner, Barcoe Tunnel and other mines in this important mining region. 110 ppgs., 10.99

The Mines of Colorado - First published in 1867, it has been unavailable since those days and sheds important light on Colorado's early mining history. Written shortly after the events took place, this publication sheds important light on the Pike's Peak Gold Rush, the discovery of gold on Ralston Creek and Dry Creek in the 1850's, as well as details on the first wave of miners into Colorado and their trials and tribulations as they crossed the Great Plains. Also included are details on early discoveries of lode gold in the mountainous regions of Colorado, details on the early mines hardrock and placer mines, and much more. It is a veritable treasure trove on Colorado's early mining history and will be of great importance to anyone who is interested in the mining of gold or other minerals in Colorado, as well as those interested in the history of the state. 478 ppgs., 29.99

The La Plata Mining District of Colorado - Originally titled "Geology and Ore Deposits in the Vicinity of the La Plata District of Colorado" and first published in 1949, it has been unavailable since those days and sheds important light on the mines and mineral deposits of the La Plata Mining District of Colorado.214 ppgs., 19.99

Washington Mining Books

The Republic Mining District of Washington - Unavailable since 1910, this important publication was originally published by the Washington Geologic Survey and has been unavailable for a century. Topics include the geology, rock formations and the formation of ore deposits in this important mining area of Washington State. Also included are hard to find details on the geology, history and locations of dozens of mines in the area. Some of the mines featured include the New Republic Mine, Ben Hur, Morning Glory, the South Republic Mine, Quilp, Surprise, Black Tail, Lone Pine, San Poil, Mountain Lion, Tom Thumb, Elcaliph and many others. **8.5" X 11", 94 ppgs, Retail Price: $10.99**

The Myers Creek and Nighthawk Mining Districts of Washington - Unavailable since 1911, this important publication was originally published by the Washington Geologic Survey and has been unavailable for a century. Topics include the geology, rock formations and the formation of ore deposits in these important mining areas of Washington State. Also included are hard to find details on the geology, history and locations of dozens of mines in the area. Some of the mines featured include the Grant Mine, Monterey, Nip and Tuck, Myers Creek, Number Nine, Neutral, Rainbow, Aztec, Crystal Butte, Apex, Butcher Boy, Molson, Mad River, Olentangy, Delate, Kelsey, Golden Chariot, Okanogan, Ohio, Forty-Ninth Parallel, Nighthawk, Favorite, Little Chopaka, Summit, Number One, California, Peerless, Caaba, Prize Group, Ruby, Mountain Sheep, Golden Zone, Rich Bar, Similkameen, Kimberly, Triune, Hiawatha, Trinity, Hornsilver, Maquae, Bellevue, Bullfrog, Palmer Lake, Ivanhoe, Copper World and many others. **8.5" X 11", 136 ppgs, Retail Price: $12.99**

The Blewett Mining District of Washington - Unavailable since 1911, this important publication was originally published by the Washington Geologic Survey and has been unavailable for a century. Topics include the geology, rock formations and the formation of ore deposits in this important mining area of Washington State. Also included are hard to find details on the geology, history and locations of dozens of mines in the area. Some of the mines featured include the Washington Meteor, Alta Vista, Pole Pick, Blinn, North Star, Golden Eagle, Tip Top, Wilder, Golden Guinea, Lucky Queen, Blue Bell, Prospect, Homestake, Lone Rock, Johnson, and others. **8.5" X 11", 134 ppgs, Retail Price: $12.99**

Silver Mining In Washington - Unavailable since 1955, this important publication was originally published by the Washington Geologic Survey. Featured are the hard to find locations and details pertaining to Washington's silver mines. **8.5" X 11", 180 ppgs, Retail Price: $15.99**

The Mines of Snohomish County Washington - Unavailable since 1942, this important publication was originally published by the Washington Geologic Survey and has been unavailable for seventy years. Featured are details on a large number of gold, silver, copper, lead and other metallic mineral mines. Included are the locations of each historic mine, along with information on the commodity produced. **8.5" X 11", 98 ppgs, Retail Price: $10.99**

The Mines of Chelan County Washington - Unavailable since 1943, this important publication was originally published by the Washington Geologic Survey and has been unavailable for seventy years. Featured are details on a large number of gold, silver, copper, lead and other metallic mineral mines. Included are the locations of each historic mine, along with information on the commodity. **8.5" X 11", 88 ppgs, Retail Price: $9.99**

Metal Mines of Washington - Unavailable since 1921, this important publication was originally published by the Washington Geologic Survey and has been unavailable for nearly ninety years. Widely considered a masterpiece on the Washington Mining Industry, "Metal Mines of Washington" sheds light on the important details of Washington's early mining years. Featured are details on hundreds of gold, silver, copper, lead and other metallic mineral mines. Included are hard to find details on the mineral resources of this state, as well as the locations of historic mines. Lavishly illustrated with maps and historic photos and complete with a glossary to explain any technical terms found in the text, this is one of the most important works on mining in the State of Washington. No prospector or miner should be without it if they are interested in mining in Washington. **8.5" X 11", 396 ppgs, Retail Price: $24.99**

Gem Stones In Washington - Unavailable since 1949, this important publication was originally published by the Washington Geologic Survey and has been unavailable since first published. Included are details on where to find naturally occurring gem stones in the State of Washington, including quartz crystal, amethyst, smoky quartz, milky quartz, agates, bloodstone, carnelian, chert, flint, jasper, onyx, petrified wood, opal, fire opal, hyalite and others. **8.5" X 11", 54 ppgs, Retail Price: $8.99**

The Covada Mining District of Washington - Unavailable since 1913, this important publication was originally published by the Washington Geologic Survey and has been unavailable for a century. Topics include the geology, rock formations and the formation of ore deposits in this important mining area of Washington State. Also included are hard to find details on the geology, history and locations of dozens of mines in the area. Some of the mines featured include the Admiral, Advance, Algonkian, Big Bug, Big Chief, Big Joker, Black Hawk, Black Tail, Black Thorn, Captain, Cherokee Strip, Colorado, Dan Patch, Dead Shot, Etta, Good Ore, Greasy Run, Great Scott, Idora, IXL, Jay Bird, Kentucky Bell, King Solomon, Laurel, Laura S, Little Jay, Meteor, Neglected, Northern Light, Old Nell, Plymouth Rock, Polaris, Quandary, Reserve, Shoo Fly, Silver Plume, Three Pines, Vernie, White Rose and dozens of others. **8.5" X 11", 114 ppgs, Retail Price: $10.99**

The Index Mining District of Washington - Unavailable since 1912, this important publication was originally published by the Washington Geologic Survey and has been unavailable for a century. Topics include the geology, rock formations and the formation of ore deposits in this important mining area of Washington State. Also included are hard to find details on the geology, history and locations of dozens of mines in the area. Some of the mines featured include the Sunset, Non-Pareil, Ethel Consolidated, Kittaning, Merchant, Homestead, Co-operative, Lost Creek, Uncle Sam, Calumet, Florence-Rae, Bitter Creek, Index Peacock, Gunn Peak, Helena, North Star, Buckeye. Copper Bell, Red Cross and others. 8.5" X 11", 114 ppgs, Retail Price: $11.99

Mining & Mineral Resources of Stevens County Washington - Unavailable since 1920, this important publication was originally published by the Washington Geologic Survey and has been unavailable for a century. Topics include the geology, rock formations and the formation of ore deposits in these important mining areas of Washington State. Also included are hard to find details on the geology, history and locations of hundreds of mines in the area. 8.5" X 11", 372 ppgs, Retail Price: $24.99

The Mines and Geology of the Loomis Quadrangle Okanogan County, Washington - Unavailable since 1972, this important publication was originally published by the Washington Geologic Survey and has been unavailable for a century. Topics include the geology, rock formations and the formation of ore deposits in this important mining area of Washington State. Also included are hard to find details on the geology, history and locations of dozens of gold, copper, silver and other mines in the area. 8.5" X 11", 150 ppgs, Retail Price: $12.99

The Conconully Mining District of Okanogan County Washington - Unavailable since 1973, this important publication was originally published by the Washington Geologic Survey and has been unavailable for a century. Topics include the geology, rock formations and the formation of ore deposits in this important mining area of Washington State, which also includes Salmon Creek, Blue Lake and Galena. Also included are hard to find details on the geology, mining history and locations of dozens of mines in the area. Some of the mines include Arlington, Fourth of July, Sonny Boy, First Thought, Last Chance, War Eagle-Peacock, Wheeler, Mohawk, Lone Star, Woo Loo Moo Loo, Keystone, Hughes, Plant-Callahan, Johnny Boy, Leuena, Gubser, John Arthur, Tough Nut, Homestake, Key and many others 8.5" X 11", 68 ppgs, Retail Price: $8.99

Wyoming Mining Books

Mining in the Laramie Basin of Wyoming - Unavailable since 1909, this publication was originally compiled by the United States Department of Interior. Also included are insights into the mineralization and other characteristics of this important mining region, especially in regards to coal, limestone, gypsum, bentonite clay, cement, sand, clay and copper. 8.5" X 11", 104 ppgs, Retail Price: $11.99

New Mexico Mining Books

The Mogollon Mining District of New Mexico - Unavailable since 1927, this important publication was originally published by the US Department of Interior and has been unavailable for 80 years. Topics include the geology, rock formations and the formation of ore deposits in this important mining area in New Mexico. Of particular focus is information on the history and production of the ore deposits in this area, their form and structure, vein filling, their paragenesis, origins and ore shoots, as well as oxidation and supergene enrichment. Also included are hard to find details, including the descriptions and locations of numerous gold, silver and other types of mines, including the Eureka, Pacific, South Alpine, Great Western, Enterprise, Buffalo, Mountain View, Floride, Gold Dust, Last Chance, Deadwood, Confidence, Maud S., Deep Down, Little Fanney, Trilby, Johnson, Alberta, Comet, Golden Eagle, Cooney, Queen, the Iron Crown, Eberle, Clifton, Andrew Jackson mine, Mascot and others. 8.5" X 11", 144 ppgs, Retail Price: $12.99

The Percha Mining District of Kingston New Mexico - Unavailable since 1883, this important publication was originally published by the Kingston Tribune and has been unavailable for over one hundred and thirty five years. Having been written during the earliest years of gold and silver mining in the Percha Mining District, unlike other books on the subject, this work offers the unique perspective of having actually been written while the early mining history of this area was still being made. In fact, the work was written so early in the development of this area that many of the notable mines in the Percha District were less than a few years old and were still being operated by their original discoverers with the same enthusiasm as when they were first located. Included are hard to find details on the very earliest gold and silver mines of this important mining district near Kingston in Sierra County, New Mexico. 8.5" X 11", 68 ppgs, Retail Price: $9.99

East Coast Mining Books

<u>The Gold Fields of the Southern Appalachians</u> - Unavailable since 1895, this important publication was originally published by the US Department of Interior and has been unavailable for nearly 120 years. Topics include the geology, rock formations and the formation of ore deposits in this important mining area of the American South. Of particular focus is information on the history and statistics of the ore deposits in this area, their form and structure and veins. Also included are details on the placer gold deposits of the region. The gold fields of the Georgian Belt, Carolinian Belt and the South Mountain Mining District of North Carolina are all treated in descriptive detail. Included are hard to find details, including the descriptions and locations of numerous gold mines in Georgia, North Carolina and elsewhere in the American South. Also included are details on the gold belts of the British Maritime Provinces and the Green Mountains. **8.5" X 11", 104 ppgs, Retail Price: $9.99**

Gold Rush Tales Series

Millions in Siskiyou County Gold - In this first volume of the "Gold Rush Tales" series, leading mining historian and editor Kerby Jackson, introduces us to the story of how millions of dollars worth of gold was discovered in Siskiyou County during the California Gold Rush. Lavishly illustrated with photos from the 19th Century, this hard to find information was first published in 1897 and sheds important light onto the gold rush era in Siskiyou County, California and the experiences of the men who dug for the gold and actually found it. **8.5" X 11", 82 ppgs, Retail Price: $9.99**

The California Rand in the Days of '49 - In this second volume of the "Gold Rush Tales" series, leading mining historian and editor Kerby Jackson, introduces us to four tales from the California Gold Rush. Lavishly illustrated with photos from the 19th Century, this hard to find information was first published in 1890's and includes the stories of "California's Rand", details about Chinese miners, how one early miner named Baker struck it rich and also the story of Alphonzo Bowers, who invented the first hydraulic gold dredge. **8.5" X 11", 54 ppgs, Retail Price: $9.99**

More Mining Books

Prospecting and Developing A Small Mine - Topics covered include the classification of varying ores, how to take a proper ore sample, the proper reduction of ore samples, alluvial sampling, how to understand geology as it is applied to prospecting and mining, prospecting procedures, methods of ore treatment, the application of drilling and blasting in a small mine and other topics that the small scale miner will find of benefit. **8.5" X 11", 112 ppgs, Retail Price: $11.99**

Timbering For Small Underground Mines - Topics covered include the selection of caps and posts, the treatment of mine timbers, how to install mine timbers, repairing damaged timbers, use of drift supports, headboards, squeeze sets, ore chute construction, mine cribbing, square set timbering methods, the use of steel and concrete sets and other topics that the small underground miner will find of benefit. This volume also includes twenty eight illustrations depicting the proper construction of mine timbering and support systems that greatly enhance the practical usability of the information contained in this small book. **8.5" X 11", 88 ppgs. Retail Price: $10.99**

Timbering and Mining - A classic mining publication on Hard Rock Mining by W.H. Storms. Unavailable since 1909, this rare publication provides an in depth look at American methods of underground mine timbering and mining methods. Topics include the selection and preservation of mine timbers, drifting and drift sets, driving in running ground, structural steel in mine workings, timbering drifts in gravel mines, timbering methods for driving shafts, positioning drill holes in shafts, timbering stations at shafts, drainage, mining large ore bodies by means of open cuts or by the "Glory Hole" system, stoping out ore in flat or low lying veins, use of the "Caving System", stoping in swelling ground, how to stope out large ore bodies, Square Set timbering on the Comstock and its modifications by California miners, the construction of ore chutes, stoping ore bodies by use of the "Block System", how to work dangerous ground, information on the "Delprat System" of stoping without mine timbers, construction and use of headframes and much more. This volume provides a reference into not only practical methods of mining and timbering that may be employed in narrow vein mining by small miners today, but also rare insights into how mines were being worked at the turn of the 19th Century. **8.5" X 11", 288 ppgs. Retail Price: $24.99**

A Study of Ore Deposits For The Practical Miner - Mining historian Kerby Jackson introduces us to a classic mining publication on ore deposits by J.P. Wallace. First published in 1908, it has been unavailable for over a century. Included are important insights into the properties of minerals and their identification, on the occurrence and origin of gold, on gold alloys, insights into gold bearing sulfides such as pyrites and arsenopyrites, on gold bearing vanadium, gold and silver tellurides, lead and mercury tellurides, on silver ores, platinum and iridium, mercury ores, copper ores, lead ores, zinc ores, iron ores, chromium ores, manganese ores, nickel ores, tin ores, tungsten ores and others. Also included are facts regarding rock forming minerals, their composition and occurrences, on igneous, sedimentary, metamorphic and intrusive rocks, as well as how they are geologically disturbed by dikes, flows and faults, as well as the effects of these geologic actions and why they are important to the miner. Written specifically with the common miner and prospector in mind, the book will help to unlock the earth's hidden wealth for you and is written in a simple and concise language that anyone can understand. **8.5" X 11", 366 ppgs. Retail Price: $24.99**

Mine Drainage - Unavailable since 1896, this rare publication provides an in depth look at American methods of underground mine drainage and mining pump systems. This volume provides a reference into not only practical methods of mining drainage that may be employed in narrow vein mining by small miners today, but also rare insights into how mines were being worked at the turn of the 19th Century. **8.5" X 11", 218 ppgs. Retail Price: $24.99**

Fire Assaying Gold, Silver and Lead Ores - Unavailable since 1907, this important publication was originally published by the Mining and Scientific Press and was designed to introduce miners and prospectors of gold, silver and lead to the art of fire assaying. Topics include the fire assaying of ores and products containing gold, silver and lead; the sampling and preparation of ore or for an assay; care of the assay office, assay furnaces; crucibles and scorifiers; assay balances; metallic ores; scorification assays; cupelling; parting' crucible assays, the roasting of ores and more. This classic provides a time honored method of assaying put forward in a clear, concise and easy to understand language that will make it a benefit to even beginners. **8.5" X 11", 96 ppgs. Retail Price: $11.99**

Methods of Mine Timbering - Originally published in 1896, this important publication on mining engineering has not been available for nearly a century. Included are rare insights into historical methods of timbering structural support that were used in underground metal mines during the California that still have a practical application for the small scale hardrock miner of today. **8.5" X 11", 94 ppgs. Retail Price: $10.99**

The Enrichment of Copper Sulfide Ores - First published in 1913, it has been unavailable for over a century. Topics include the definition and types of ore enrichment, the oxidation of copper ores, the precipitation of metallic sulfides. Also included are the results of dozens of lab experiments pertaining to the enrichment of sulfide ores that will be of interest to the practical hard rock mine operator in his efforts to release the metallic bounty from his mine's ore. **8.5" X 11", 92 ppgs. Retail Price: $9.99**

A Study of Magmatic Sulfide Ores - Unavailable since 1914, this rare publication provides an in depth look at magmatic sulfide ores. Some of the topics included are the definition and classification of magmatic ores, descriptions of some magmatic sulfide ore deposits known at the time of publication including copper and nickel bearing pyrrohitic ore bodies, chalcopyrite-bornite deposits, pyritic deposits, magnetite-ileminite deposits, chromite deposits and magmatic iron ore deposits. Also included are details on how to recognize these types of ore deposits while prospecting for valuable hardrock minerals. **8.5" X 11", 138 ppgs. Retail Price: $11.99**

The Cyanide Process of Gold Recovery - Unavailable since 1894 and released under the name "The Cyanide Process: Its Practical Application and Economical Results", this rare publication provides an in depth look at the early use of cyanide leaching for gold recovery from hardrock mine ores. This volume provides a reference into the early development and use of cyanide leaching to recover gold. **8.5" X 11", 162 ppgs. Retail Price: $14.99**

California Gold Milling Practices - Unavailable since 1895 and released under the name "California Gold Practices", this rare publication provides an in depth look at early methods of milling used to reduce gold ores in California during the late 19th century. This volume provides a reference into the early development and use of milling equipment during the earliest years of the California Gold Rush up to the age of the Industrial Revolution. Much of the information still applies today and will be of use to small scale miners engaging in hardrock mining. **8.5" X 11", 104 ppgs. Retail Price: $10.99**

Leaching Gold and Silver Ores With The Plattner and Kiss Processes - Mining historian Kerby Jackson introduces us to a classic mining publication on the evaluation and examination of mines and prospects by C.H. Aaron. First published in 1881, it has been unavailable for over a century and sheds important light on the leaching of gold and silver ores with the Plattner and Kiss processes. **8.5" X 11", 204 ppgs. Retail Price: $15.99**

The Metallurgy of Lead and the Desilverization of Base Bullion - First published in 1896, it has been unavailable for over a century and sheds important light on the the recovery of silver from lead based ores. Some of the topics include the properties of lead and some of its compounds, lead ores such as galenite, anglesite, cerussite and others, the distribution of lead ores throughout the United States and the sampling and assaying of lead ores. Also covered is the metallurgical treatment of lead ores, as well as the desilverization of lead by the Pattinson Process and the Parkes Process. Hofman's text has long been considered one of the most important early works on the recovery of silver from lead based ores. **8.5" X 11", 452 ppgs. Retail Price: $29.99**

Ore Sampling For Small Scale Miners - First published in 1916, it has been unavailable for over a century and sheds important light on historic methods of ore sampling in hardrock mines. Topics include how to take correct ore samples and the conditions that affect sampling, such as their subdivision and uniformity. Particular detail is given to methods of hand sampling ore bodies by grab sample, pipe sample and coning, as well as sampling by mechanical methods. Also given are insights into the screening, drying and grinding processes to achieve the most consistent sample results and much more. **8.5" X 11", 124 ppgs. Retail Price: $12.99**

The Extraction of Silver, Copper and Tin from Ores - First published in 1896, it has been unavailable for over a century and sheds important light on how historic miners recovered silver, copper and tin from their mining operations. The book is split into three sections, including a discussion on the Lixiviation of Silver Ores, the mining and treatment of copper ores as practiced at Tharsis, Spain and the smelting of tin as it was practiced by metallurgists at Pulo Brani, Singapore. Also included is an overview and analysis of these historic metal recovery methods that will be of benefit to those interested in the extraction of silver, copper and tin from small mines. **8.5" X 11", 118 ppgs. Retail Price: $14.99**

The Roasting of Gold and Silver Ores - First published in 1880, it has been unavailable for over a century and sheds important light on how historic miners recovered gold and silver rom their mining operations. Topics include details on the most important silver and free milling gold ores, methods of desulphurization of ores, methods of deoxidation, the chlorination of ores, methods and details on roasting gold and silver ores, notes on furnaces and more. Also included are details on numerous methods of gold and silver recovery, including the Ottokar Hofman's Process, the Patera Process, Kiss Process, Augustin Process, Ziervogel Process and others. **8.5" X 11", 178 ppgs. Retail Price: $19.99**

The Examination of Mines and Prospects - First published in 1912, it has been unavailable for over a century and sheds important light on how to examine and evaluate hardrock mines, prospects and lode mining claims. Sections include Mining Examinations, Structural Geology, Structural Features of Ore Deposits, Primary Ores and their Distribution, Types of Primary Ore Deposits, Primary Ore Shoots, The Primary Alteration of Wall Rocks, Alterations by Surface Agencies, Residual Ores and their Distribution, Secondary Ores and Ore Shoots and Vein Outcrops. This hard to find information is a must for those who are interested in owning a mine or who already own a lode mining claim and wish to succeed at quartz mining. **8.5" X 11", 250 ppgs. Retail Price: $19.99**

Garnets: Their Mining, Milling and Utilization - First published in 1925, it has been unavailable since those days and sheds important light on the mining, milling and utilization of garnets. Included are details on the characteristics of garnets, where they are found and how they were mined. 78 ppgs, 10.99

Gemstones and Precious Stones of North America - Leading mining historian Kerby Jackson introduces us to a classic mining publication on the gems and precious stones of the United States, Canada and mexico. First published in 1890, it has been unavailable since those days and sheds important light on the gems and precious stones that may be found in North America. Included are chapters on diamonds, corundum, sapphire, ruby, topaz, emerald, disapore, spinel, turquoise, tourmaline, garnets, beyrl, peridot, zircon, quartz crystals, feldspars, pearls and many others. Included are details on where these gems and precious stones may be found throughout North America, as well as their characteristics. 360 ppgs, 24.99

Mining Camps and Mining Districts - First released in 1885 by Charles Howard Shinn under the title "Mining Camps: A Study in American Frontier Government", this publication offers a unique look at how early gold miners established their own forms of representative government during the California Gold Rush. Drawing on the the early mining codes of mideviel German miners in the Harz Mountains, on the mining customs of the Cornish tin miners and early Spanish mining laws introduced into California, the miners established the first governments in the American West. 340 ppgs, 24.99

BLM Field Handbook for Mineral Examiners - Leading mining historian Kerby Jackson introduces us to a classic mining publication on mine evaluation. First published in 1962, this work sheds important light on the techniques of BLM Mineral Examiners to perform validity on mining claims. 132 ppgs, 10.99

<u>**Six Months In The Gold Mines During The California Gold Rush**</u> - Unavailable since 1850, this important work is a first hand account of one "49'ers" personal experience during the great California Gold Rush, shedding important light on one of the most exciting periods in the history of not only California, but also the world. Compiled from journals written between 1847 and 1849 by E. Gould Buffum, a native of New York, "Six Months In The Gold Mines During The California Gold Rush" offers a rare look into the day to day lives of the people who came to California to work in her gold mines when the state was still a great frontier. **8.5" X 11", 290 ppgs. Retail Price: $19.99**

<u>**The Discovery of Gold in Australia**</u> - First published in 1852, it has been unavailable since those days and sheds important light on Australia's gold mining history. Included are rare communications between British agents and the British Crown when gold was first discovered in Australia in 1851. This rare text contains hard to find details on Australia's first mining camps and Britain's early attempts to provide for the orderly regulation of gold mines in that part of the world. Also of interest are hard to find extracts of articles that appeared in the early colonial newspapers that did their best to report on Australia's gold rush as it took place.
102 ppgs, 10.99

www.ingramcontent.com/pod-product-compliance
Lightning Source LLC
Chambersburg PA
CBHW080717190526
45169CB00006B/2411